The End of Ownership

The Information Society Series

Laura DeNardis and Michael Zimmer, series editors

Interfaces on Trial 2.0, Jonathan Band and Masanobu Katoh

Opening Standards: The Global Politics of Interoperability, Laura DeNardis, editor

The Reputation Society: How Online Opinions Are Reshaping the Offline World, Hassan Masum and Mark Tovey, editors

The Digital Rights Movement: The Role of Technology in Subverting Digital Copyright, Hector Postigo

Technologies of Choice? ICTs, Development, and the Capabilities Approach, Dorothea Kleine

Pirate Politics: The New Information Policy Contests, Patrick Burkart

After Access: The Mobile Internet and Inclusion in the Developing World, Jonathan Donner

The World Made Meme: Public Conversations and Participatory Media, Ryan Milner

The End of Ownership: Personal Property in the Digital Economy, Aaron Perzanowski and Jason Schultz

The End of Ownership

Personal Property in the Digital Economy

Aaron Perzanowski and Jason Schultz

The MIT Press
Cambridge, Massachusetts
London, England

First MIT Press paperback edition, 2018

This book was set in Stone Sans and Stone Serif by Toppan Best-set Premedia Limited. Printed and bound in the United States of America.

Library of Congress Cataloging-in-Publication Data

Names: Perzanowski, Aaron, author. | Schultz, Jason M., author.
Title: The end of ownership : personal property in the digital economy / Aaron Perzanowski and Jason Schultz.
Description: Cambridge, MA : The MIT Press, 2016. | Series: The information society series | Includes bibliographical references and index.
Identifiers: LCCN 2016013180 | ISBN 9780262035019 (hardcover : alk. paper)—9780262535243 (pbk.)
Subjects: LCSH: Personal property. | Internet—Law and legislation. | Electronic commerce—Law and legislation. | Intellectual property.

Classification: LCC K783 .P47 2016 | DDC 346.04/8—dc23 LC record available at http://lccn.loc.gov/2016013180

10 9 8 7 6 5 4 3 2

For Clem, who at least gets my record collection—AP
For Kate & Elliott, my two favorite booksneaks—JS

Contents

Acknowledgments

This project grew out of a series of our academic articles: "Digital Exhaustion," *UCLA Law Review* 58 (2011): 889–946; "Copyright Exhaustion and the Personal Use Dilemma," *Minnesota Law Review* 96 (2012): 2067–2143; "Legislating Digital Exhaustion," *Berkeley Technology Law Journal* 28 (2015): 1535–1557; and "Reconciling Personal and Intellectual Property," *Notre Dame Law Review* 90 (2015): 1213–1263. It also draws from "What We Buy When We *Buy Now*," by Aaron Perzanowski and Chris Jay Hoofnagle, forthcoming in the *University of Pennsylvania Law Review* (2016). We thank those journals, their editors, and our many colleagues who helped refine our thinking at those early stages. We'd also like to extend our appreciation to the long list of law schools where those articles and portions of this book were workshopped.

We are grateful to Peter Steffensen for outstanding research assistance and to Yawen Li for last-minute editing. For their thoughtful comments on this manuscript, we'd like to thank Brandon Butler, Ryan Calo, Rochelle Dreyfuss, Dave Fagundes, Josh Fairfield, Tomás Gómez-Arostegui, James Grimmelmann, Parker Higgins, Chris Hoofnagle, Peter Jaszi, Ariel Katz, Mark Lemley, Lydia Loren, Mike Madison, Christina Mulligan, Sean O'Connor, John Rothchild, Sarah Schindler, Matthew Schruers, Sherwin Siy, Molly Shaffer Van Houweling, and Fred von Lohmann.

1 Introduction

Chances are, you are reading this book in one of two ways. Either you are holding a bound set of printed pages—a traditional analog book—or you are holding an electronic device displaying a digital file—an ebook. Whether the page is physical or virtual, the words are the same. But the seemingly simple choice between these two ways of delivering text offers a window into a broader set of questions about the emerging digital economy and our place within it. In the courts, in the marketplace, and in our homes, we find mounting evidence that our rights to own, control, repair, and use the products we buy depend, in large part, on whether those goods are analog or digital. This looming rift between buyers of analog and digital goods is the byproduct of a number of relatively recent legal, technological, and marketplace developments. And those shifts implicate not only media content like books, music, and movies, but also nearly every software-enabled device we encounter, from phones, cars, and coffeemakers to medical devices like pacemakers and insulin pumps.

An example may help illustrate the problem. In George Orwell's dystopian classic *1984*, the Ministry of Truth, at the behest of Big Brother, destroyed documents by casting them into the memory hole, a massive network of tubes leading to an incinerator. Amazon, the world's largest bookseller, sells *1984*—along with millions of other titles—both in print and in its Kindle ebook store. Assuming they had a chance to read the book first, Kindle users were no doubt struck by the irony of Amazon's decision to remotely delete their purchased copies of *1984* in response to a dispute with a publisher.[1] These customers went to bed one night thinking they owned a copy of Orwell's cautionary tale and woke up the next morning to find their book had been confiscated. In its place, they received a refund and an object lesson in the risks of digital reading.

In the world of printed books, this scenario would be unthinkable. Your local bookseller cannot creep into your home in the middle of the night and reclaim the contents of your bookshelf. But Amazon exercises a very different kind of practical power over your digital library. Your Kindle runs software written by Amazon, and it features a persistent network connection. That means Amazon can send it instructions—to delete a book or even replace it with a new version—without any intervention from you.

But it's not just the technology that sets analog and digital books apart. The legal terrain looks very different as well. If you bought a printed copy of this book, it became your personal property. Like your favorite pair of shoes, or your toothbrush, you own it. Ownership of this book means you can do lots of things with it. You can keep it forever; you can read it as many times as you like; you can lend it to a friend;[2] you can resell it or give it away; you can leave it to a loved one in your will. We don't encourage it, but you can even burn it if you feel like it. Because of the demands of copyright law, you generally cannot make copies of this book without permission. But otherwise, if you own it, it is yours to do with as you choose. This may seem obvious; the same basic rules of personal property have applied to books and other movable property for hundreds of years.

And you might expect digital books to work much the same as their printed counterparts. They contain the same text and are often sold by the very same retailers for comparable prices. Indeed, a 2012 study showed that nearly a third of bestselling ebooks were more expensive than their hardcover counterparts.[3] But according to publishers and retailers, ebooks play by a distinct sets of rules. For print, we rely on the familiar rules of personal property. But do you actually own your ebooks? Most readers have probably never paused to ask this question. After all, you clicked the "Buy Now" button and paid the price demanded by your favorite ebook retailer. Why wouldn't you own the thing you bought?

Despite the common sense appeal of that view, digital retailers insist that ownership depends on the terms of an end user license agreement ("EULA")—that incomprehensible slew of legalese you reflexively click "I agree" to dismiss. Those terms—negotiated by lawyers working for retailers and publishers—determine your rights, not the default entitlements of personal property. And buried within those thousands of words that we all ignore is one consistent message: you don't own the books you bought; you merely license them. That is to say, you have permission to read them. Until one day, you don't.

The *1984* incident is hardly the only case of readers losing access to their purchases. Linn Nygaard, a Norwegian Kindle customer, lost dozens of ebooks she bought from Amazon. They simply vanished without notice when Amazon erased her Kindle, citing unspecified "abuse of [its] policies."[4] Our best guess is that Nygaard ran afoul of those policies because she lived in Norway, a territory in which Amazon had not yet launched its Kindle Store. But we can't say for sure, since Amazon never bothered to tell her. To be clear, she didn't pay with a stolen credit card; she didn't hack Amazon's servers to get her ebooks for free; she simply made her purchases from the wrong country. After a worldwide spate of critical news coverage, Amazon relented and restored Nygaard's purchases. But Amazon's technical ability and legal authority to take away your ebooks remain unchanged.

Other retailers have caused ebook purchases to vanish without even the pretense of wrongdoing on the part of readers. Scholastic, the publisher of children's educational books, launched its Storia ebook platform in 2012, promising that purchases could be shared with up to ten students. But just two years later, Scholastic announced a change of plans. It would be offering ebooks exclusively through a streaming model. And its new subscription service required an active Internet connection. No Wi-Fi—the reality in too many of the underfunded schools across the United States—means no reading.[5] Subscription services are not inherently bad. They can offer those of us interested in temporary access real value, but Scholastic's approach retroactively converted what students and educators thought were purchases into rentals—from permanent possession to conditional permission. As the publisher explained, "The switch to streaming means that eBooks you've previously purchased may soon no longer be accessible."[6]

The chapters that follow will illustrate that this problem goes well beyond ebooks. Digital distribution of music has already largely displaced CD sales. And digital movie distribution is projected to overtake DVD and Blu-ray within the next few years. Software and video game sales are trending toward digital models as well. In each of these sectors, the same story about ownership plays out. Gamers who buy titles on discs can lend them to friends and resell them. Those who download their games through Xbox Live or the PlayStation Network can do neither, even though they pay the same price. The rights that we have come to expect when we buy music, movies, and other content are at best uncertain and at worst absent in the digital marketplace.

So how did our rights in media goods become so unstable and insecure? Part of the answer is technology. Cheap remote storage, high-speed mobile network connections, and nearly ubiquitous computing devices like tablets and smartphones have facilitated new ways of distributing media. Digital downloads, cloud storage, and streaming services offer convenience, instant accessibility, and lower prices to consumers. But they also physically separate us from the books we read, the music we play, and the movies we watch. That content doesn't live on our shelves anymore. It's in a server farm in some distant and unknown city.

At the same time, aggressive intellectual property laws, restrictive contractual provisions, and technological locks have weakened end user control over the digital goods we acquire. We will tackle each of these developments in detail later, but the terms of use for the Kindle Store offer a brief glimpse into one chief cause of the instability consumers confront in the digital marketplace. As Amazon explains in the EULA you've likely never read, "Kindle Content is licensed, not sold, to you."[7] In other words, you don't own the ebooks that you buy. What's more, "if you fail to comply with any term of this Agreement, ... Amazon may immediately revoke your access to ... Kindle Content without refund."[8] So if you break Amazon's rules by, for example, posting a "threatening, defamatory, ... or objectionable" product review, your books can be confiscated.[9] Your rights are defined by a nonnegotiable agreement you've never read, one that—as we will show—runs counter to what most of us think we can do with the products we buy.

Beyond these contractual restrictions, many products today incorporate technology that restricts how you can use them. Digital rights management (DRM) builds these restrictions into the very design of the products we buy. If you've ever found yourself unable to watch a movie because you've authorized too many devices, you've been the victim of DRM. But DRM isn't limited to digital media. Today, we see it in all manner of products, where it plays much the same role, officiously telling you what you can and can't do with the stuff you buy. When Keurig released version 2.0 of its home coffee machine in 2014, for example, it incorporated DRM to prop up sales of its coffee. Customers who tried to brew cheaper, off-brand ground coffee were greeted by a message on the device's display that politely refused to make their cup of coffee, instructing them to buy Keurig-brand coffee instead.

The traitorous coffee maker is not an isolated example. The same trends that threaten to undermine ownership of intangible digital media have made their way into the world of tangible objects. Smartphones, televisions, cars, household appliances, and wearable technology like the Apple Watch and Fitbit—to name just a few—feature embedded software and network connectivity that control how we use the things we buy. And like Kindle ebooks, the agreements that accompany these products typically insist that buyers are merely licensed to use them and expressly prohibit lending, resale, modification, and even repair.

You might find this vision of the future troubling. But the anemic understanding of consumer rights that manufacturers and retailers are pushing is only one side of the story. Ownership is a contested question, and the digital marketplace is a contested space. As we will argue, there are good reasons to resist these efforts to redefine our relationship to the media and devices that shape so much of our interaction with the world. And while some courts and policymakers have been led down the path of ever-diminishing consumer rights, others have signaled an unwillingness to jettison those rights without carefully considering the consequences.[10] Perhaps more importantly, readers, listeners, and tinkerers—everyday people—are expressing their own reluctance to accept ownership as an artifact of some bygone predigital era. The questions we address in this book are complex, and there are no easy answers. Our goal is to explain the current state of our relationship with the products we buy, how we arrived at this pivotal moment in ownership's history, and to begin what we hope is an open and ongoing conversation about where we might go from here.

Of course, any discussion of our digital future has to acknowledge the benefits of new technologies and the business models they enable. Many of us—including the authors of this book—embrace the digital marketplace. Just consider how the Kindle revolutionized the experience of reading. Today's devices can store thousands of books in a package smaller and lighter than the average paperback. They allow readers to search, bookmark, and annotate, to share favorite passages with a community of friends, and to instantly define unfamiliar words. And new books are a mere click away thanks to wireless connectivity and integrated shopping platforms. Even those of us who prefer the reassuring heft of a hardcover, the smell of ink on paper, and afternoons wandering the aisles of the Strand, Powell's, or John K. King can at least recognize the appeal of digital books.

Beyond books, many of us happily store our collections of digital artifacts in the cloud. Or we opt for no permanent collections at all, instead dipping into the streams of all-you-can-eat subscription content available from Netflix, Spotify, and the like. As the popularity of these streaming services makes clear, lots of us are content to sacrifice ownership and permanence if it means a wider selection, more portability, greater convenience, and lower prices. Advocates of licensing models say they enable a degree of flexibility that sales simply can't. If customers can license the precise rights that meet their needs—to read a book, but not lend it, or to watch a movie on your smartphone, but not on your TV—they can pay accordingly, and everyone wins. We will return to price discrimination—the notion of charging different customers different prices depending on their specific preferences and willingness to pay. For now, it's enough to say that we agree that certain forms of price discrimination increase consumer choice in valuable ways. But we think the benefits of price discrimination are often overstated, and that it can do more harm than good if unrestrained.

Today, we operate in a market that—for the most part—affords a choice between ownership and more conditional, impermanent access to digital and physical goods. Those choices are neither right nor wrong. But they have consequences, both for individuals and society more broadly. There are things we gain and things we lose. And if we know what those trade-offs are, we can make more informed, more meaningful choices—not only about the products we buy, but also about the laws and policies that govern the marketplace.

So what is at stake when we make these choices? The most immediate consequence of nonownership is the long list of substantive rights we lose. The prohibitions found in most EULAs and enforced by most DRM contrast starkly with the default rules of private property. You can't resell a product you don't own. You can't lend it, give it away, or donate it. You can't read, watch, or listen on unapproved devices. You can't modify or repair the devices you use. There might be good reasons to give up those rights. But the evidence we will present strongly suggests that most consumers are poorly informed about the disparities between ownership and licensing.

Nor is the impact of the shift from ownership to licensing limited to individuals; our educational and cultural institutions are dealing with the fallout as well. When a library buys a printed book, for example, it can lend it to as many patrons as it chooses, without asking the publisher for permission or paying any additional fees. Library books can remain in circulation for decades, serving the needs of hundreds of readers. But when libraries acquire ebooks, licensing terms and software code often impose hard ceilings on lending. HarperCollins ebooks, for example, can be lent out twenty-six times, which translates to a single year of borrowing, after which they essentially self-destruct.[11] Patrons cannot borrow that title again until the library ponies up an additional fee to the publisher. So despite the claims by publishers like Random House—who claim that libraries "own" their ebooks—libraries don't own their digital collections any more than you own the movies on your Netflix queue.[12]

Digital consumers sacrifice stability and permanence too. As the *1984* episode shows, purchases can be deleted or disabled without warning or explanation for any number of reasons. Perhaps you unknowingly violated some provisions of a site's terms of service. Perhaps the retailer adopted a new business model that left existing customers in the cold. Google, Major League Baseball, MSN Music, Sony, Virgin Digital, Walmart, and Yahoo all pulled variants of this move when they decided to shut down the servers that customers needed to access the media they purchased.[13] Some customers were given the chance to convert to other services, but many were told to burn their purchases to CDs or lose them forever. In other cases, retailers have simply gone out of business altogether. Although you might lament your local bookstore closing up shop, you'd at least keep your books. But when HDGiants, a purveyor of high-quality audio and video files went bankrupt, its servers went dark and its paying customers were left with nothing.[14]

Privacy presents another concern.[15] For analog media, we have strong privacy protections that limit access to information about what books you check out from the local library and what movies you rent from your local Redbox. Putting the law aside, practical barriers ensured that governments, publishers, and retailers could not easily track who bought, owned, resold, or enjoyed analog copies of banned and confiscated works like *Tropic of Cancer,*[16] *As Nasty As They Wanna Be,*[17] or *The Tin Drum.*[18] Digital transactions make this kind of tracking far easier. First, digital purchases are almost always tied to a unique user account, linking your purchase history to your identity. Second, the architecture of online media allows unprecedented surveillance of consumer behavior. Adobe, for example, recently came under fire when researchers discovered that its popular ebook platform, Digital Editions, reported back not only the titles of every book in a reader's library, but also when they were read and even what pages were viewed. Even more troubling, this information was sent over the Internet unencrypted, meaning that any mildly sophisticated hacker could learn all there is to know about your reading habits.[19] And then of course, there is the risk of government surveillance by the National Security Agency (NSA) and others.[20]

The transition from owning to licensing causes another, more widespread problem. Because their terms can vary so widely, licenses lead to uncertainty about what rights we actually acquire. When it comes to ownership, centuries of practice—reinforced by clear legal rules—mean that when a reader walks into a store and exchanges cash for a book, they know with a fair degree of certainty what they are getting.[21] That clarity disappears when rights are defined by the variable and often incomprehensible text of a license agreement. Licenses vary—from retailer to retailer, from publisher to publisher, from product to product. Close study of the license accompanying an Amazon ebook tells you very little about one you might buy from Apple. And it tells you almost nothing about the license for your coffeemaker. Licenses are driven by the concerns of manufacturers, retailers, and publishers—and the negotiations among them. As a result, licenses are often idiosyncratic and subject to change, sometimes even after your purchase. The rights you acquire are therefore less clear and less predictable than the rights associated with ownership.

Beyond its impact on individuals, this erosion of clarity poses a risk of broader social harms. One advantage of clear, reliable property rights is that they make it easier for people to navigate the marketplace. Replacing clear property rules with complicated and uncertain contractual ones makes life harder for all of us and impairs the functioning of the economy as a whole.

In the language of economists, property rights increase efficiency by lowering transaction costs. Transaction costs are all of the costs aside from the sticker price that we incur when we buy a product or engage in some transaction.[22] Let's say you want to buy a newly released bestseller. The retail price for the book is $25. But that price doesn't take into account all of the relevant costs of acquiring the book. You have to drive to the bookstore; you have to spend time looking for the book on the shelf; in some cultures, you may have to haggle over the price. These are all transaction costs. Even information about the book comes at a cost. We have to investigate products to determine their quality and characteristics before deciding to buy them. How many reviews, for example, did you read before deciding to buy this book?

Clear property rights help keep these costs low.[23] Without stable and reliable rules about what rights we acquire when we buy a product, information costs go up. On the one hand, when you see the price tag on a book, you understand that if you pay the $25, you own it. And most of us have a solid understanding of what ownership entails. On the other hand, in a world where some books were owned, some could be read only once, others had to be returned after a month, and still others could be read in the bathtub but not on the beach, you'd need to carefully investigate each purchase. You'd have to ask the sales associate lots of questions or scour the terms that accompany each book to figure out precisely what rights you acquire, for what period of time, and what restrictions apply.

This information cost problem leads to what economists call an externality—a cost created by a transaction that isn't borne by the parties striking the deal. Pollution is a classic example.[24] A factory makes widgets and sells them to the public. In the process, the factory emits pollution that lowers air quality. The price of the resulting widgets is a function of a number of factors—the cost of labor, materials, research and development, and advertising, among others. But the cost of pollution isn't one of them. Widget buyers don't pay for it, and in the absence of some environmental regulation, the factory doesn't either. So pollution is a cost created by the sale of the widget that neither the buyer nor the seller has to take into account.

Information costs can work the same way. Let's say your neighbor loves to read at the beach, but prefers a quiet glass of bourbon in the bathtub. So they are enthusiastic about the prospect of saving a dollar on their next book by paying for the beach-but-no-bathtub license. You, on the other hand, prefer to own your books. When your neighbor and others like them opt for the licensed book—assuming they are fully informed about their choice—they may be getting precisely what they want. As between buyer and seller, this deal looks like a success. But there is a cost they are both ignoring. The next time you go to the bookstore, you'll have to keep a careful eye out for licensed books, lest you end up drawing a bath only to find out you are prohibited from reading. So information costs for you and other would-be book owners increase. The fact that some books come with idiosyncratic rules imposes a cost on all book shoppers, regardless of their preferences.

This isn't the only externality created by the shift away from ownership. There are other costs that go unnoticed in our calculations. One benefit of ownership is preservation. Valuable cultural works disappear for all sorts of reasons. Government censorship can remove works from the market; books and records go out of print when they are deemed commercially unviable; films—from *The Interview* to Disney's *Song of the South*—are hidden from view for reasons that range from political controversies to pure marketing ploys.[25] Works can also be lost to accidents, natural disasters, and plain old inattention. Ownership helps guard against those losses. When we own our copies, we have greater incentives to make efforts to preserve them, and it's harder for publishers and government actors to erase them. And when works are distributed widely on secondary markets through resale and lending, the risk of loss is reduced. Even though we all benefit from the preservation of our shared cultural heritage, outside of the small circle of archivists and cultural historians, few of us give it much thought. So when we choose to license rather than own, we are—in admittedly small increments—chipping away at preservation efforts.

Ownership can also spur innovation. When the used goods we buy can be resold, those secondary markets create an incentive for new and improved products. We see new features on our cars and phones, remastered music, and behind-the-scenes features for movies, in part because ownership and transferability increase competitive pressure. This trend is perhaps most visible in the video game industry where publishers frequently release "Game of the Year" or other special editions loaded with extra content as a way to compete with cheaper used copies. Ownership also enables user innovation from those who modify and improve the products they buy.[26] This innovation is valuable. To the extent licensing reduces incentives and opportunities for innovation, it imposes costs on society that are not reflected in the lower price of licensed goods. Precisely because these costs are not felt acutely by individuals, we might doubt whether consumer choice alone—even if informed—can fully solve the problems licensing creates around information costs, preservation, and innovation.

Competition can also get a boost from individual ownership because it helps lower the costs of switching from one format, device, or platform to another. Lower switching costs open the market up to new entrants with potentially superior products. Imagine you are a loyal Microsoft Xbox enthusiast with thousands of dollars invested in hardware and software. But you're considering switching sides and buying a Sony PlayStation. If you own your Xbox, you can sell it along with your collection of games on Craigslist or eBay. But if Microsoft could stop you from reselling your device and games—as it currently does for digital games purchased through its Xbox Live service—you'd be less inclined to switch, and the market would be less competitive as a result.

But the most fundamental value at stake in the choice between ownership and licensing is autonomy—the sense of self-direction, that our behaviors reflect our own preferences and choices rather than the dictates of some external authority. If we own our purchases, we are free to make whatever lawful use of them we choose. If you own your books, you can give them away. If you own your records, you can lend one to a friend. If you own your iPhone, you can use the mobile carrier and install the apps of your choice. If you own your PlayStation, you can replace its operating system and use it as a low-cost computer. If you own your Ferrari, you can customize it as you see fit. And if you own your Keurig coffeemaker, you can brew whatever brand of coffee you prefer. What ties these disparate behaviors together is that they don't depend on permission. You don't have to ask Amazon or Apple or Sony. You are free to act on your own accord, even over their objections.

That's one reason we find efforts to recreate resale, lending, and other rights through licensing unsatisfying. Amazon created a program, for example, that allows readers to "lend" an ebook to a friend. But that program has strings attached. An ebook can only be lent a total of one time and only for fourteen days. Most crucially, lending depends on permission from the book's publisher. As a result, only a small fraction of ebook titles allows lending. Your hardcovers, on the other hand, can be lent to as many friends, relatives, or strangers as you choose whether the publisher likes it or not. So while Amazon has recreated some aspects of the lending culture we have grown accustomed to for print books, digital lending remains an imperfect simulacrum, in large part because it hinges on choices other than our own.

Of course even with ownership, we don't enjoy total freedom. There are limits on what we can do with the things we own. But those limits are generally defined by law. And under our system, law is created through a process—imperfect in many respects—that is ultimately responsive to our input. But a future defined by licensing is one where control over how we interact with the world around us and with each other is increasingly concentrated in the hands of a small coterie of powerful private actors. In that future, the limits on our autonomy will flow from a EULA rather than collective self-government. It doesn't have to be this way. Technology can constrain our freedom, but it can also empower us.

In 1984, the U.S. Supreme Court weighed the fate of the VCR. Movie studios sued Sony, alleging that TV viewers used its Betamax player to unlawfully record broadcast programs. Ultimately, the Court rejected this effort to dictate how new technologies were designed and used by their owners. Key testimony in the case came from an unlikely source, Fred Rogers—host of the PBS mainstay *Mister Rogers' Neighborhood*. In characteristically simple and powerful language, he explained the value of the VCR in terms of personal autonomy: "I have always felt that with the advent of all of this new technology that allows people to tape the *Neighborhood* off-the-air ... they then become much more active in the programming of their family's television life. Very frankly, I am opposed to people being programmed by others. My whole approach in broadcasting has always been 'You are an important person just the way you are. You can make healthy decisions.' ... Anything that allows a person to be more active in the control of his or her life, in a healthy way, is important."[27] Ownership facilitates the sort of active participation that Mister Rogers had in mind. And the licensing model puts it at risk.

So far, we've focused on how ownership affects the average person. But there is another set of interests at stake in this debate. Much of the effort to displace ownership has been undertaken in the name of strengthening the intellectual property (IP) rights of creators. Intellectual property is generally understood as a way for the law to provide economic incentives for the creation of new inventions and works of expression. By protecting inventors and authors from copying, IP law boosts their chances of financial success.

And if the theory behind IP protection is correct, we see more creativity as a result.

IP rights holders—from publishers to carmakers—are attracted to the increased control licensing promises them. They can eliminate secondary markets like used book stores; they can reduce competition for complementary products like coffee or ink cartridges; and they can corner the market for repair and other related services. All of which, they argue, increases their incentives to invest in new and better products. Moreover, rights holders argue that digital goods are fundamentally different from analog ones. They can be copied perfectly and distributed at no cost. Unlike a paperback that falls apart after a handful of readings, an ebook can be passed around to infinite readers. We agree that analog and digital goods are not perfect substitutes, though we think the differences between them are often overstated. Still, we acknowledge that the rules of digital ownership can't simply copy and paste from the analog world. But we shouldn't simply scrap ownership either.

If greater IP protection comes at the cost of personal property rights, a licensing-only strategy may well backfire. Today, most commercially valuable copyrighted works are available for free somewhere online, with or without the copyright holder's permission. The challenge facing copyright law—and with the introduction of 3D printing, soon patent law too—is figuring out how to convince the public to pay for things it can get for free. One way the law does that is through the stick of infringement liability. And that stick is a big one. The Copyright Act allows for damages of up to $150,000 for unlawfully downloading a single song.[28] But copyright holders, despite their best efforts, cannot locate and sue each and every downloader on the Pirate Bay.[29] And the probability of a lawsuit is too low to deter many of them.

If we want to persuade people to pay for these products rather than download them illegally, the carrot can be just as important as the stick. People pay for things that offer them good value for their money. And ownership is a major component of that value. Property rights mean that buyers have assurances about their ability to use and enjoy the products they buy. A book, movie, or video game that the law recognizes as your personal property is more valuable than one in which you have no recognized rights. And as a result, personal property rights provide a strong reason to buy lawful copies. But when copies lack the rights and freedoms we expect, they are less desirable and harder to distinguish from free, infringing ones.

The risk is that creators, publishers, and digital retailers are unwittingly reducing incentives to buy their own products through aggressive efforts to control how readers, listeners, and viewers use them. If after learning about the restrictions they impose, people are not convinced that digital products present a good value proposition, we can expect a number of responses. Some will revert back to analog copies, if they can. Others will decide to spend their money on subscription services like Spotify and Netflix, which are arguably less profitable for copyright holders than sales-based business models. Some will choose to download content illegally. And some will decide to spend their disposable income elsewhere, on a vacation or personal trainer, for example. Tampering with ownership is likely to have major consequences, and perhaps not the ones creators expect.

That's our argument for why these issues—and this book—matter. Here is how the remaining chapters will proceed. First, we outline some basic principles of personal and intellectual property law—in particular, the notion of exhaustion of rights—to lay the conceptual groundwork for the rest of the book. Next, we trace two key developments in the erosion of ownership—the technologies of digital distribution and the rise of the license agreement. Then, we explore the mismatch between the fine print of EULAs and the claims about "buying" and "owning" that are so prevalent in the digital marketplace. We will demonstrate that those claims mislead consumers about the fundamental nature of digital transactions. From there, we turn our attention from individuals to the implications of the licensing model for an important group of institutional actors, public libraries. Next, we look at how the licensing model, which was largely confined to digital media for decades, has been exported to the world of physical goods. That transition starts with DRM technology and the laws that protect it. But with the emergence of the Internet of Things, the question of our relationship with the devices around us—and sometimes in us—is more pressing than ever. Then we explore another legal avenue for exerting control over how we use the objects we buy—the patent system—and how the ongoing fight over so-called post-sale restrictions threatens ownership. Finally, we will outline an agenda to reconcile stable, reliable personal property rights with our inevitably digital future.

2 Property and the Exhaustion Principle

In order to make sense of our changing relationship with digital goods, we need to start with a basic understanding of our system of property rights. This chapter should make a few things clear. For one, defining consumer rights through licenses rather than the default rules of ownership is a significant departure from the way we typically treat personal property. For another, despite common misconceptions, property rights are rarely absolute. Instead, they often have to accommodate the interests of others. The law has developed ways of resolving those competing claims. Who wins and who loses in those struggles for control over valuable resources tells us something about our priorities, about what sorts of uses and what sorts of users we think should be privileged under the law. When it comes to goods subject to intellectual property protection, the principle of exhaustion is the primary tool for resolving disputes between IP holders and personal property owners. The shift away from exhaustion and toward licensing is an effort to take power away from individuals in favor of copyright holders and their retail partners. Because this power grab reduces efficiency, creates harmful externalities, and interferes with individual autonomy, we should find it troubling.

A Property Law Primer

To begin, we should distinguish between four basic types of property: real, personal, intellectual, and intangible. Most of us associate the term "property" with land—your home or the proverbial family farm. In the law, we call this real property, as in real estate. Real property stands apart from other kinds of property in a number of ways. First, each piece of real property, defined by the physical space it occupies, is unique. Theoretically, each parcel of land can be clearly defined and distinguished from all others. Second, because real property is tied to physical space, it stays put. Setting aside things like tectonic shifts, landslides, and rivers changing course, your real property will remain exactly where you left it. Third, real property is comparatively expensive. As a result, most of us engage in only a handful of real property transactions in our lives. Compare the number of homes you've bought to the number of socks, books, or cell phones you've purchased.

Another key distinguishing characteristic of real property is the relative flexibility the law affords owners of land to define and rearrange their rights. Interests in real property can take a number of forms. The most familiar is what lawyers call the fee simple. The owner of a fee simple interest has the right to use the land, to possess it, to exclude others, to sell it, to give it away, and to collect profits from it. But there are other ways to own real property. A life estate, for example, is a right to possess and use land, but only for the duration of a person's life. Tenancy in common allows two or more owners to possess land at the same time, each having equal rights to occupy and use the property. Timeshares, in contrast, make it possible for several owners to use the property, but each for only a set amount of time each year. And condominiums provide for individual ownership of each unit in a building, but joint ownership of common areas like lobbies and hallways.[1]

Not only can real property owners choose from these and other ready-made forms of ownership, they can further customize their property rights using legal tools called real covenants and equitable servitudes.[2] With these tools, property owners can craft limitations or obligations on future use of the property; they can attach strings that dictate what can be done with the property. These strings are said to "run with the land," binding all subsequent owners of the property. Through these legal devices, owners can impose a wide range of restrictions on future generations. They can limit a piece of land to a particular purpose—a single family home, for example. Or they can forbid certain uses—no lighthouses or organic supermarkets. They can prohibit pets. They can insist on green lawns and gardeners to tend them. They can require that you choose from an approved palette of paint colors. They can ban holiday decorations. In short, through property law, owners can fashion their own bespoke sets of rights, imposing their preferences and whims on everyone who encounters a particular parcel of real property.[3]

When it comes to personal property, the rules are not nearly as customizable.[4] Property interests in chattels—legal-speak for your personal possessions—are far more prosaic. You can own a tuxedo; you can rent a tuxedo; and you can borrow a tuxedo. But the law of property doesn't recognize timeshares in tuxedos.[5] Nor does the law recognize servitudes on chattels;

personal property doesn't come with strings attached. Although some English courts in the mid-1800s toyed with the idea of applying servitudes to movable property, they quickly corrected course. And U.S. courts followed suit.[6] As a result, your tuxedo can't be burdened by an obligation to wear a particular brand of shoes or a prohibition on wearing it two weekends in a row. Of course, you could agree to those limitations in a contract, but only the parties to a contract are bound by its terms. Those obligations would not "run with the tuxedo" to bind future owners.

Maintaining clear and simple rules for personal property serves a couple of related purposes. First, it helps keep information costs in check. We may be willing to carefully investigate each real estate transaction for idiosyncratic property obligations. After all, we don't buy homes all that often, and a lot of money is at stake. But when it comes to donuts, staplers, and books, such effort hardly seems worth it. The cost of determining the precise contours of the property rights in these sorts of purchases could easily exceed the value of the good itself.[7] Second, clear property rights make sure that common items of everyday commerce can be freely bought and sold. If tuxedo manufacturers could customize the rights that buyers obtain, they might be tempted to protect themselves from competition by prohibiting rental or controlling resale prices. As one court put it, efforts to attach strings to personal property are "obnoxious to public policy, which is best subserved by great freedom of traffic in such things as pass from hand to hand."[8]

Our personal property rules place a high value on alienability—the right of an owner of an item to resell it, give it away, or otherwise transfer it. But it doesn't have to be that way. We can imagine other property systems that would lead to very different outcomes. Consider goblin property. In J. K. Rowling's *Harry Potter* series, goblins are skilled metalsmiths. And they are deeply attached to the items they craft, regarding themselves as the true owners of those items, even after their sale. As Rowling explains: "Goblin notions of ownership, payment, and repayment are not the same as human ones. ... To a goblin, the rightful and true master of any object is the maker, not the purchaser. All goblin-made objects are, in goblin eyes, rightfully theirs. ... They would consider it rented by [a purchaser]. ... They consider our habit of keeping goblin-made objects, passing them from wizard to wizard without further payment, little more than theft."[9]

But we are not goblins, at least not yet. When you sell your used Prius, you don't owe Toyota a percentage of the sale price. When you die, Samsung doesn't get to reclaim your TV. Once you've purchased an item, it belongs to you.

So far we've considered property interests that relate to physical assets. With intellectual property, our focus shifts from the physical to the incorporeal. IP encompasses a cluster of legislative and judicial rules that grant property-like rights in the intangible creations of human ingenuity. Patent law rewards inventors with exclusive rights in their novel innovations; copyright law provides rights to creators of original expressive works; and trademark law protects distinctive symbols from confusingly similar uses. These and other related legal regimes fall under the broad umbrella of intellectual property.

Although patents will play an important role later in our story, our primary focus is copyright law.[10] Copyright is concerned with works of creative expression. Books, music, film, visual art, and software all fall within its broad subject matter. In order for a work to qualify for copyright protection, it must be original—it must reflect a modicum of creativity, and it can't be copied from an existing work.[11] To be protected by copyright, a work must also be recorded in some physical form.[12] This fixation requirement is satisfied when a writer jots down a story in a notebook, or when a photographer saves an image to a memory card.

If a work qualifies, the copyright holder is granted a number of valuable exclusive rights. They include copying the work, selling or otherwise transferring copies of the work, publicly performing or displaying the work, and making new works based on it.[13] Only the copyright holder is legally entitled to engage in these behaviors without permission. To take a more concrete example, let's say you buy a copy of a film on Blu-ray. Unless you have permission from the copyright holder, you can't make copies of that film; you can't play it for a roomful of strangers; and you can't make an unauthorized sequel. This ability to control how the work is used by others accounts for the property-like nature of copyright interests. Since copyrighted works are often sold in physical form, IP rights enable a degree of ongoing control by copyright holders over the tangible products we buy. But as we will describe later in this chapter, that control is constrained in crucial ways by the principle of exhaustion.

Although IP shares some characteristics with more familiar forms of property, it differs from them in a number of respects. In fact, many question whether the term "intellectual property," despite its wide usage, overstates the connection between IP and other more familiar forms of property. Unlike most forms of tangible property, patents and copyrights expire. Indeed, the U.S. Constitution requires that they last for only "limited times." Initially, the term was fourteen years. Today, patents expire after twenty years, and copyrights expire an astounding seventy years after the death of the author.[14]

More fundamentally, the mental creations of interest to IP law are what we call public goods; they exhibit two characteristics that set them apart from traditional property. First, ideas and expression are nonrivalrous. Use of an intellectual resource by one person doesn't interfere with its use by another. If I am driving my car, you can't. And no matter how hard you try, you can't fit two people into a single tuxedo. But millions of people can watch the same television show, sing the same song, or read the same novel without depleting the underlying intellectual resource. As Thomas Jefferson explained it, "He who receives an idea from me, receives instruction himself without lessening mine; as he who lights his taper at mine, receives light without darkening me."[15] Second, information is nonexcludable; it is difficult to maintain control of intellectual resources once they have been disclosed. You can put a fence around your land to keep out intruders. You can lock your jewels in a safe. But controlling the use and spread of information is like trying to construct a fence around your own personal supply of air. For these two reasons, intellectual resources are distinguishable from other kinds of property.

Good ideas—like a new way of treating a deadly disease or the perfect breakup song—have the potential to improve lives. We want them to spread. So we should celebrate the fact that information goods, unlike arable land or iPhones, don't run out or wear down. But the public-goods characteristics of information resources create a potential problem. Although a groundbreaking treatment or a heartbreaking song can be freely shared and enjoyed, making something new requires investments of time, effort, and money. If creators cannot recover those investments, plus a reasonable profit for their trouble, some will be dissuaded. In a world where creating new works is expensive and copying them is cheap and easy for the public, poets will become accountants, and inventors will become plumbers. IP law is meant to remedy this public goods problem—the feared undersupply of creative investment—by creating legal barriers to competition by prohibiting copying. IP rights are an effort to overcome the inherent characteristics of intellectual resources and force them to behave more like tangible property.

Importantly, not all intangible resources fall under the IP umbrella. Interests in debts, securities, and government franchises—think liquor licenses or taxi medallions—all concern intangible assets rather than tangible objects, but they aren't regulated by IP law. Likewise, we can think of assets like digital currencies and virtual objects—a powerful weapon in your favorite video game, for example—in terms of property. The rules surrounding these relatively new intangible assets remain largely undefined.[16] Digital objects don't easily fit into either the IP or the personal property frameworks. Consider a digital movie you purchase from Apple. You browse on iTunes, find a movie that looks promising—we recommend Shane Carruth's *Upstream Color*—and buy it for $12.99. You can stream that movie to your TV from Apple's servers, you can download it to your laptop, or you can come back to it another day. But what set of rules defines your rights in that digital asset? Is it the rules of personal property or the rules of intellectual property and, by extension, the iTunes license agreement? The question at the heart of this book is whether digital goods—both media content and devices with embedded software—play by the familiar rules of personal property or the more flexible but often opaque rules of IP licenses. In short, do we own our digital goods?

Understanding Ownership

What does it mean to own property? That's a surprisingly hard question to answer. Legal scholars, economists, and philosophers have debated the fundamental nature of property for centuries. And we can't hope to button up that long-running dialogue here. Our more modest goal is to convince you to reconsider some of your own preconceptions about ownership.

Most people think of property ownership as bestowing an absolute right to an individual owner of a tangible thing. If you know a single adage about the law of property, it is probably William Blackstone's oft-quoted reference to property as "that sole and despotic dominion which one man claims and exercises over the external things of the world, in total exclusion of the right of any other individual in the universe."[17] But our discussion already suggests what Blackstone knew very well: that this absolutist view of property is an oversimplification.[18]

Over the last century or so, a more nuanced understanding has gained prominence among property experts. Under this view, property ownership is comprised of a bundle of distinct and separable interests. Consider the owner of a piece of land. We can break the concept of ownership down into a number of discrete rights that the owner might enjoy: the right to possess the land, to have physical control over it; the right to use the land, for picnics or kite flying for example; the right to manage the land, to decide who else can have picnics and fly kites; the right to income from the land, to charge rent for picnics and kite flying; and the right to alienate the land, to sell it or give it away. Each of these rights, among others, contributes to the owner's property interests. But no single right is essential to ownership. The owner could lease their land, but their lack of possession doesn't mean they no longer own it. They could sell mineral rights in the land to an energy company, but the company's use and profit are not inconsistent with individual ownership.

This understanding of property as a bundle of related but separable rights helps us account for the complexity and flexibility of property interests, especially interests in land. It is a less useful metaphor when it comes to personal property, where the law insists on less complex arrangements of rights in order to limit information costs, ensure the free movement of goods, and protect consumer welfare. When it comes to your tuxedo, the law doesn't permit this sort of slicing and dicing. But copyrights, in contrast, lend themselves to the bundle-of-rights model by congressional design. The rights of a copyright holder are an explicitly enumerated bundle—to reproduce, distribute, publicly display and perform, and make adaptations of the work. And each of those rights is divisible by time, geography, medium, or any other limitation dreamed up by a rights holder. A playwright, for example, can transfer the right to publicly perform their latest play but, if they so choose, only on alternating Tuesdays in Wisconsin. In practice, this can easily lead to dozens of owners of copyright interests in a single work. In theory, the number is infinite.

These two conceptions of property—one emphasizing simplicity, the other embracing flexibility—are in tension. On the whole, we think the bundle-of-rights view is a more accurate description of how property works. But we also recognize the importance of placing limitations on the flexibility that the bundle enables. Particularly when it comes to consumer goods—characterized by their low cost and high volume compared to real property—we think the information costs and other negative externalities that flow from customized bundles of rights make the case for a limited menu of standard transactions.

Let's say you decide to take us up on our recommendation of *Upstream Color*. For the sake of simplicity, let's also stipulate that you prefer hard copies. You have four basic options. You can buy the film on Blu-ray. You can rent it from Redbox or a local video rental store, if you can find one. You can get it from a subscription service like Netflix's vestigial DVD-by-mail service. Or you can borrow it from a friend or the local library. That short list of familiar transactions would satisfy the needs of most people.

We could come up with other fanciful alternatives. They might even end up looking something like the Apple or Amazon EULAs. Taken to its extreme, the bundle-of-rights view would allow for the creation of any bespoke assemblage of rights and restrictions we can imagine. And while there might be some people who prefer these innovative transactions, we think that degree of flexibility imposes costs on individuals and society without a corresponding increase in consumer satisfaction. Flexibility is valuable, but only up to this point of diminishing returns.

The second widely held belief we want to challenge concerns the subject matter of property rights. Most of us think about property as conferring rights over things, typically tangible things. But we've already seen how the law extends some property concepts to intangibles. So if property rights don't define an owner's relationship to an object, what do they define? For many legal scholars, the answer is that property rights actually define relationships between people. To say I own a parcel of land, or a tuxedo, or a song is to say that I have the power to control, to varying degrees, your behavior in connection with the thing I own. I don't control the thing, but the ways in which others interact with it.

From this perspective, property law is just one of many tools that allow us to structure our relationships with others and influence their behavior. Contracts are another. But again, contractual rights are only enforceable against the parties to an agreement. Property rights in contrast don't require negotiation, mutual agreement, or assent. Your property rights apply to everyone, whether they like it or not. Property and contract can also differ in the sorts of remedies they provide. For example, infringement of a copyright can lead to tens or even hundreds of thousands of dollars in damages, regardless of any measurable harm suffered by the copyright holder. But a claim based purely on a contract would be limited to actual provable losses caused by the breach.

Third, we should say a bit about the source of property rights. Some see property as a natural right—one that exists independently of any legal rule and rooted in some deeper philosophical foundation. John Locke famously argued that property is a natural right that arises out of labor. According to Locke, we acquire property rights by exerting effort to gather resources, cultivate land, or develop new ideas.[19] Hegel offered another view on the foundations of property. He argued that property is necessary for individual self-actualization. Unless we can exert control over objects in the world, we cannot express our will, achieve our goals, or thrive as individuals.[20]

But regardless of its foundation, property is deeply contingent on the law as a practical matter. Property rights exist by virtue of the willingness of the government to recognize and ultimately enforce them. Imagine you encounter trespassers on your land. What do you do? Perhaps you call the police. Under some circumstances, you might file a lawsuit. If the legal system refuses to deal with the trespassers, your property rights have very little value. Even a property owner's self-help remedies depend on legal recognition. Physically removing someone from your property is lawful only to the extent the legal system favors your interests over those of the trespassers.

IP rights are even more obviously contingent on legal recognition. Copyright and patent rights in the United States simply wouldn't exist without legislation. There is no recognized natural right to such protection. Instead, the law protects authors and inventors in order to promote the creation of valuable intellectual resources for society at large to use and enjoy. And Congress and the courts tweak those property interests—adding new rights, expanding or limiting existing ones—when they are persuaded that changes would yield a better outcome.

The existence and specific contours of property rights are dictated by our legal system. So by calling something property and calling someone its owner, the law is using a sort of shorthand. Those labels identify the winner of a contest for control over the behavior of others with respect to some valuable resource. At this point, our claim about the future of ownership should start to take on a somewhat clearer meaning. When we say that personal property rights are being eroded or eliminated in the digital marketplace, we mean that rights to use, to control, to keep, and to transfer purchases—physical and digital—are being plucked from the bundle of rights purchasers have historically enjoyed and given instead to IP rights holders. That in turn means that those rights holders are given greater control over how each of us consume media, use our devices, interact with our friends and family, spend our money, and live our lives. Cast in these terms, it is clear that there is a looming conflict between the respective rights of consumers and IP rights holders. The next question is how do we resolve it?

Property Conflicts

In the public imagination, property rights definitively resolve conflicts in favor of property owners. But that's not how property actually works. What happens when your interests are at odds with those of your neighbor? Or more pointedly, what happens when the personal property interests of purchasers are at odds with the intellectual property rights of copyright holders? Property rights—of all varieties—are limited in their scope. They have baked-in constraints that prevent owners from disregarding the interests of others. In real property, eminent domain and nuisance ordinances are useful reminders that the power of property owners is finite. Title to your property will not excuse excessive noise or pollution that harms your neighbors, for example. Similarly, personal property owners have to comply with all sorts of generally applicable laws that restrain use of their property. You can't park your car on the sidewalk or swing your favorite ax in a crowded park, even though you own them.

IP rights feature their own inherent limits. Copyright, for example, extends only to the particular expression used by an author, not to the ideas that underlie their work. So the copyright in *Star Wars* protects the film, script, plot, and even specific characters from copying, but it does not give George Lucas, or now Disney, exclusive rights to the hero's journey.[21] And the fair use doctrine sometimes permits copying of the author's expression if it serves the public interest and poses limited risk of economic harm to the copyright holder. For example, our copying of a short quote from *Harry Potter and the Deathly Hallows* to illustrate a point about property law a few pages back is a fair use that doesn't require permission from the copyright holder.

Conflicts between the property interests of copyright holders and consumers are central to our story. For creators, intellectual property law is primarily concerned with intangible creations—who owns them, how to exploit them, and whether one creation treads too closely to another. But for most of us, copyright law is a set of rules that tells us what we can and can't do with our stuff. Can you copy your Blu-ray movies to your laptop? Can you share a favorite new album with a friend? Can you sell your used books? How many people can you invite into your home to watch the Super Bowl? And if you watch it at the local bar, how big can the TV be?[22] In this sense, copyright law constrains how we use our property on a daily basis.

So there is a tug of war going on between purchasers and IP holders. If the law strengthens IP rights, it narrows personal property rights. And if the law gives us more latitude to do what we want with the things we buy, IP holders sacrifice some control over us. This tension is an inevitable feature of a system that accounts for the interests of both creators and consumers. From a policy perspective, intellectual and personal property rights function as stand-ins for a broad range of concerns—creative incentives, information costs and other externalities, deception, and autonomy. Since the law creates and defines the contours of those rights, the question of how best to balance them is inevitable. For well over a century, copyright law has provided a transparent and predictable answer to that question through the principle of exhaustion. That principle is the legal backdrop against which the rest of the developments we describe take place.

The Exhaustion Principle

Exhaustion is the notion that an IP rights holder relinquishes some control over a product once it sells or gives that product to a new owner. We say those IP rights have been exhausted because the rights holder can no longer control many of the uses the new owner may make of that product. The power to prevent distributing, displaying, and sometimes reproducing a work gives way to the personal property interests of the owners. This principle is expressed in a number of copyright rules, the most important of which is the first sale doctrine. The Copyright Act prohibits unauthorized distribution—the selling, renting, leasing, or giving away—of protected works. Without some exception or limitation, we would have no right to donate our used books or sell our used video games or even give a newly purchased CD to a friend on their birthday.[23] The first sale doctrine steps in to prevent that absurd result. It allows copy owners to sell, give away, lend, or rent their copies even when the copyright holder objects.

Here's one recent example of the first sale doctrine at work. In 1982, Atari released *E.T. the Extra-Terrestrial*, a video game based on the hit movie, for its 2600 home console. The game bombed and today is widely regarded as the worst video game ever made. In 1983, Atari shipped as many as twenty semitrucks loaded with unsold copies of the game to Alamogordo, New Mexico, where they were promptly buried. Atari reportedly chose that particular landfill because no scavenging was allowed. It wanted to erase all evidence of this embarrassing creative misstep. Decades later, the city decided the games were worth more above ground than below it. The city dug them up and auctioned off nine hundred surviving *E.T.* cartridges for more than $100,000.[24]

But first sale isn't copyright's only exhaustion rule. The Copyright Act also prohibits unauthorized public displays of protected works. Again, without some exception, that means that a museum that paid millions of dollars for a painting would need the copyright holder's permission before hanging it on the wall. Luckily, the Act makes clear that the owner of a painting or other work is free to display it.[25] Another provision gives owners of copies of software the right to make copies necessary to run, back up, or modify it.[26] And long before exhaustion was written into the Copyright Act, courts recognized the rights of owners of copies of books and other works to resell them, to copy portions of them, and to create new works from them. What all of these exhaustion rules have in common is that they privilege the property interests of copy owners over those of copyright holders.

Exhaustion is copyright law's main tool for mediating the tension between intellectual and personal property rights. For more than a century, this cluster of exhaustion-based rules struck a balance that gave purchasers and other owners considerable, but not unlimited, rights to use and enjoy their copies. But at the same time, those rules have helped protect the financial interests of copyright holders in two ways. First, they limit the rights that flow from exhaustion to owners of lawful copies of a work. So if the copy you bought is an infringing one, exhaustion would not entitle you to resell it. Second, even an owner of a copy can't do whatever they want with it. For example, you can't buy a copy of the latest bestselling young adult novel and make copies for all of your friends. Exhaustion doesn't go that far. Nor should it.

Exhaustion is so deeply engrained in our experience of the copyright economy and such a fundamental part of our property rules that most of us barely notice it at work. But that doesn't mean it isn't important. Exhaustion is the reason we have used record stores and bookstores. It's the reason we have public libraries and eBay. It's the reason we can lend a novel to a friend and leave our record collections to loved ones in our wills. It's the reason museums can display their paintings and you can back up your software. But the rules that permit these uses are not a given. They were established by the courts and Congress, and their survival depends on continued legal recognition.

The first cases in the United States to recognize the exhaustion principle date back to the nineteenth century. For example, Mark Twain's plan to market *The Adventures of Huckleberry Finn* exclusively through a high-priced subscription service was thwarted when book distributors sold copies to bookstores. When Twain sued the distributors, the court reasoned that since they owned the books they bought from Twain, the distributors were free to sell them to whoever they chose.[27] Subsequent courts went even further. They decided that fire-damaged pages of books sold as wastepaper could be bound and resold over the objections of copyright holders;[28] that booksellers could repair and resell used copies of children's schoolbooks, even if that meant reproducing missing or damaged parts;[29] and that the purchaser of loose pages of Rudyard Kipling poems could bind them together with other works to create a new collection.[30] Although courts were not unanimous in embracing this burgeoning exhaustion principle, the majority agreed that ownership of a copy entitled the owner to make a variety of uses that otherwise would have been illegal.

Eventually, copyright exhaustion made its way to the Supreme Court in a case that pitted a publisher against the retailer Macy's.[31] In 1904, the

Bobbs-Merrill Company published the novel *The Castaway*. Like many publishers at the time—and today—Bobbs-Merrill was deeply interested in controlling retail prices of its books. In an effort to inflate those prices, Bobbs-Merrill printed the following notice, an early ancestor of today's EULAs, in each copy of *The Castaway*: "The price of this book is one dollar net. No dealer is licensed to sell it at a less price, and a sale at a less price will be treated as an infringement of the copyright."

Macy's sold copies of the book for a mere eighty-nine cents, and Bobbs-Merrill promptly sued for copyright infringement. Bobbs-Merrill argued that since it had the right to choose whether to sell the book to the public or not, it could sell it with conditions attached that would bind all subsequent buyers. But the Supreme Court recognized this theory as the literary equivalent of servitude on a tuxedo. And it wasn't buying it. According to the justices, once Bobbs-Merrill sold copies at its chosen wholesale price, its right to control the further distribution of those copies came to an end. Copyright law does not recognize that sort of ongoing control over the personal property of another.

Almost immediately, Congress embraced the first sale rule. It passed the Copyright Act of 1909 just a year later, which provided that "nothing in this Act shall be deemed to forbid, prevent, or restrict the transfer of any copy of a copyrighted work the possession of which has been lawfully obtained."[32] And when Congress overhauled the Copyright Act in 1976, it retained the first sale rule in a slightly modified form that remains in effect today.[33] Because *Bobbs-Merrill Co. v. Straus* was concerned with resale, the rule recognized by the Court, and subsequently by Congress, was silent on the kinds of copying and alteration endorsed in earlier decisions. Nonetheless, the exhaustion principle was firmly established.

For well over a century now, exhaustion has been the law of the land in the United States. And it has proven good public policy. Individuals and society more broadly benefit from rules that allow owners to exercise property rights in their purchases of copyrighted materials. By opening up secondary markets, the exhaustion principle promotes access to cultural works. More people can read books, watch films, and play games when used copies, rentals, and lending drive down the cost of access. Exhaustion is also central to securing the benefits of privacy, preservation, innovation, and competition that flow from consumer ownership of products that contain copyrighted works.

And exhaustion—as an example of a clear property rule—helps keep information costs in check. The rules of exhaustion are simple, intuitive, and familiar. They largely track the rules that apply to other forms of tangible property. As a result, we don't have to engage in painstaking research to determine our rights in each book, movie, or game we buy. We already know the rules. That makes life easier and markets more efficient. From the perspective of the average person, exhaustion is an easy sell.

Resisting Exhaustion

But copyright holders have resisted exhaustion at nearly every turn. Many seem to regard it as some sort of loophole that allows owners to make uses that should require permission or additional payment. Although these misgivings about exhaustion have been around for decades, rights holders have started to take particularly aggressive steps to circumvent exhaustion and weaken consumer property interests in recent years.

The efforts of book publishers to restrain the free alienability of private property were responsible for the explicit recognition of the exhaustion doctrine. The restrictions Bobbs-Merrill tried to impose on owners of copies of *The Castaway* were themselves a rejection of the idea that buyers of products can control whether and at what price they can be resold. Bobbs-Merrill, of course, lost that battle. But one hundred years later, publishers are still filing copyright suits in an effort to control resale prices for books.

In 2008, John Wiley & Sons, a multibillion-dollar publisher of college textbooks, sued a USC graduate student for reselling textbooks on eBay. Supap Kirtsaeng emigrated to the United States from his native Thailand to study math. For those of you who haven't set foot in a college bookstore recently, a single required textbook can cost as much as $300. But as Kirtsaeng understood, publishers sell those same books overseas at far more sensible rates, sometimes as much as 90 percent lower. Knowing an opportunity when he saw one, Kirtsaeng began importing books bought in foreign markets and selling them online to eager students at U.S. universities. This looks like just the kind of free alienability that the exhaustion principle is meant to enable.

But Wiley saw things differently. It argued that because these books were printed outside of the United States, the first sale doctrine didn't apply, and therefore Kirtsaeng's resales were illegal. In 2013, the Supreme Court rejected that contention.[34] The Court held if the books were sold by the publisher, exhaustion applied regardless of where they were manufactured. In part, the Court was worried that to rule otherwise could mean that any product manufactured overseas that included a copyrighted work couldn't be sold, rented, or given away without permission.[35] Those works would include not only the novel you bought on vacation, but also your smartphone and your car—devices that integrate copyrighted software essential to their operation.

John Wiley & Sons and its supporters warn that preventing international price discrimination—the practice of charging high prices in the United States and lower prices in developing countries—will ultimately harm the most vulnerable readers. Publishers like Wiley say that if they can't segment markets, they will be forced to either raise prices in countries like Thailand or stop selling products there altogether. But there are good reasons to be skeptical about these threats. First, publishers can prevent imports without eliminating first sale. They could make books in developing economies available through rental or subscription models, for example. Second, the impact of market segmentation on price is a thorny question with no clear answer. Pricing decisions involve a number of variables that make predictions and generalizations difficult.[36] So while the net global effect of exhaustion is likely to reduce cost, its impact in any given geographic market is hard to predict. For every country like Thailand that benefits from low book prices, there's one like South Africa, where even before *Kirtsaeng* an overwhelming number of citizens reported that books were too expensive to buy.[37] Or consider India, where academic publishers tend to supply outdated editions, and the latest versions cost just as much as they do in Western countries.[38]

There's another strategy that publishers like Wiley should consider—admittedly, one they won't like. They could make less money. The average U.S. college student spends about $900 a year on textbooks.[39] A recent Bureau of Labor Statistics report reveals that textbook prices have gone up over 1,000 percent in the past three decades. That's three times more than the increases for medical services, new home prices, and the consumer price index.[40] Not surprisingly, textbook publishers have been highly profitable. McGraw-Hill's profit margin in 2012 was 25 percent; Wiley's was 15 percent.[41] The year before, the margins for Wiley's scientific, medical, technical and scholarly division were a staggering 42 percent.[42] In comparison, "Big Oil" sees profits around 5 percent, and Walmart's hover near 3 percent.[43]

Given the gouging of U.S. students, the college textbook market was ripe for an enterprising importer like Kirtsaeng.[44] If publishers lowered their prices in the United States, demand for imports would decrease significantly. That might lower publisher profits. But we are confident they could make do. Copyright law is designed to ensure sufficient incentives for the production of new works, but it should not be a license to print money. As the Supreme Court has explained, "Creative work is to be encouraged and rewarded, but private motivation must ultimately serve the cause of promoting broad public availability of literature, music, and the other arts."[45]

Book publishers aren't the only ones to push back against exhaustion. In the early 1990s, record labels worried that used CDs would harm their bottom line. CDs, after all, were different from vinyl records or cassettes since digital copies don't deteriorate with age or use. Or so the worry went.[46] So the major labels were openly hostile to used CD sales and plotted ultimately unsuccessful strategies to undermine the practice. They tried to crack down on stores that sold used CDs by refusing returns on opened merchandise and withholding millions of dollars of customary underwriting for print and radio ads.[47] They even threatened to boycott stores that sold used CDs, turning away orders for the new album from then-megastar Garth Brooks.[48]

Hollywood demonstrated its own antipathy to exhaustion. In the late 1970s, when the home video market was first emerging, videocassettes of Hollywood films were designated as "sold for home use only" in an effort to dissuade both public performance and rental. In fact, contracts with retailers explicitly prohibited rental of the tapes they purchased. Video rental pioneer George Atkinson faced mixed signals from the movie industry as he tried to build his business. Finally, after receiving a legal threat for renting tapes he purchased lawfully, Atkinson consulted a lawyer who informed him that thanks to the first sale rule, he was free to rent the tapes he owned. Although the studios required retailers to promise not to rent tapes and refused to sell directly to rental shops, the free alienability of personal property ensured a steady inventory for Atkinson and his contemporaries,[49] leading major studios like Warner Brothers, Disney, and MGM to lobby for the doctrine's repeal.[50]

More recently, a video producer of a different sort has adopted a firmly anti-exhaustion posture. Beachbody, maker of the popular P90X home workout videos, insists that its customers do not own the DVDs they purchase from the company's website, but merely license these physical purchases.[51] Beachbody has aggressively targeted individuals who resold legitimate copies of its DVDs on eBay, threatening litigation and demanding exorbitant compensation. It is easy to understand why Beachbody would want to prevent customers from reselling their workout videos after their New Year's resolve runs out. As reasonable as three easy payments of $39.95 may be, used DVDs could decrease sales and put pressure on the company to lower prices. What's harder to see is how this strategy can be squared with the principle of exhaustion, personal property rights, or the best interest of consumers.

The computer software industry has been waging its own war on exhaustion practically since its inception. In the next chapter, we will discuss the licensing practices of software companies in greater detail. For now, it's enough to note that the industry pioneered the widespread use of license agreements as a strategy for undermining end user ownership and exhaustion. Other innovations in the software industry, from digital rights management to the software-as-a-service business model, have helped developers put even greater distance between software transactions and the traditional rules of private property. Legislatively, the industry successfully lobbied Congress to pass the Computer Software Rental Amendments Act, which prevents the rental of most software programs.[52]

Today, the video game industry is the fiercest opponent of exhaustion. Its relationship with exhaustion has been contentious since at least the 1980s. Unlike other kinds of software, Congress did not carve out an exception to exhaustion that prevented rentals of console games. Despite the clear legality of video game rentals, companies like Nintendo saw rentals as a threat to game sales and characterized the rental shops as copyright infringers. Nintendo even sued Blockbuster. But because of the first sale doctrine, the game giant was forced to settle for claims that Blockbuster unlawfully photocopied the instruction manuals packaged with its games.[53] Although game rentals continue through subscription services like Gamefly, the industry has turned its focus to what it deems a bigger economic threat: used game sales.

At $2 billion per year, the used game market represents a significant portion of gaming industry revenue. Video game retailer GameStop has been the leader in this space, but used gaming has attracted the attention of Amazon and Walmart among others. Leading game developers have called the used market a "bigger threat than piracy."[54] Others have prophesied the industry's undoing if gamers continue to resell their $60 video games after finishing them. But resale has killed gaming the same way it killed the market for new music, movies, cars, and sofas. That is to say, it hasn't. In fact, GameStop reports that 70 percent of trade-in value—the money gamers get for selling their used titles—goes to the purchase of new games. That's well over $1 billion per year.[55]

Nonetheless, in response to the used game threat—real, imagined, or invented—the gaming industry has been hard at work on strategies to reduce or eliminate used sales. Most controversially, console makers have developed technologies to prevent the use of pre-owned games. Sony filed a patent application on technology that would tie individual game discs to particular users or consoles, though it hasn't yet deployed it.[56] And when

Microsoft initially announced its Xbox One console, it unveiled plans to restrict secondhand games. But after persistent and overwhelmingly negative feedback from customers, Microsoft was forced to relent.[57] As we will discuss in the next chapter, however, the gaming industry's most effective gambit against exhaustion has been the shift to digital game distribution. As one developer put it, "digital distribution stabs the used games market in the heart."[58]

If the used market directs so much revenue back into new game sales, why are publishers trying to snuff it out? One possibility is that they overestimate the losses due to used games. Not every used game translates into a lost sale. Without the used market, some gamers would simply buy fewer titles or none at all. This is the same faulty logic that led record labels to overstate the impact of file sharing a decade ago.[59] Another answer is that copyright holders are not always particularly skilled at recognizing the potential value of markets they don't control. After Hollywood lost its legal battle against the VCR, home video became a bigger source of revenue than the box office. And the music industry, after years of resisting, was dragged kicking and screaming into the era of digital distribution only after Napster threatened CD sales. Since then, Apple alone has sold over thirty-five billion songs.[60] A third explanation has less to do with miscalculation and more to do with principle. Again, the movie studios' attitude toward the VCR is instructive. When pressed in *Universal City Studios, Inc. v. Sony Corporation of America* to identify how the VCR hurt their bottom line, the studios admitted it didn't cause "a great deal of harm." Instead, their chief concern was "a point of important philosophy that transcends even commercial judgment." Their worry was that the VCR crossed "invisible boundaries" and that copyright holders "lost control."[61]

We think this focus on absolute control is shortsighted. There are good reasons to think exhaustion helps game makers, just as home video helped movie studios. Exhaustion broadens markets and expands audiences. Used games drive demand for consoles and build the community of gamers. Today's used game buyers—once they finish school and get a job—might very well start purchasing new games. The used game market also trains gamers to pay someone—even if it isn't the publisher—for their games rather than getting them for free. And it exposes them to new titles and publishers, potentially creating valuable lifelong fans in an era of long-running game sequels. These are all reasons to doubt that the used market is harmful to publishers on the whole.

Regardless, many copyright holders still see exhaustion and, by extension, personal property rights as an unfortunate legal loophole to be closed at the first opportunity. And really, what business wouldn't eliminate competition if it could? Ford dealers would happily ban used car lots and Craigslist ads. Levi's would do away with vintage shops, garage sales, and sewing machines. And restaurants would declare leftovers contraband. But of course, we'd never let them. Nor should we let copyright holders eliminate resale and lending. Exhaustion and the personal property rights it recognizes are an inherent part of copyright law's balance between the rights of creators and the rights of the public.

Of course, that is not to say that the particular balance exhaustion has struck in the past is a perfect fit for the digital economy. The exact contours of those rules, what rights they reserve for consumers and what rights they grant copyright holders, are not set in stone. That balance can and should adapt over time in response to changing conditions, like the emergence of digital distribution that we describe in chapter 3. But there are good reasons our legal system has recognized the personal property interests of consumers who buy tangible goods. And there are good reasons to retain the basic framework of personal property—one that allows for flexibility but places limits on customization—when it comes to digital goods. The standard menu of transactions—buy, rent, borrow, and give—and the default rules of ownership serve the needs of readers, viewers, and users pretty well. But a system that allows licenses to redefine those standard transactions in whatever way best serves the interests of rights holders imposes real costs. So we favor exhaustion not because *property* is a label with talismanic properties, but because it is smart policy. The basic principle of exhaustion—the notion that owners have rights that are not contingent on copyright holder permission—can and should survive the transition to a digital copyright economy. Rights holders have always fought against this principle, but the digital marketplace gives them their best chance to kill it.

3 Copies, Clouds, and Streams

In little more than a decade, the way we acquire copyrighted content has been transformed three times. Until the early 2000s, we mostly interacted with books, music, and movies as physical objects we could hold in our hands. Sure, we watched movies in theaters and listened to music on the radio, but copyright holders were primarily in the business of selling tangible copies, and legal digital downloads were still mostly hypothetical. It wasn't until Apple's iTunes Music Store launched in 2003 that a viable authorized digital distribution system emerged. This shift from tangible copies to digital ones posed major challenges for copyright law that it hasn't yet fully resolved.

But the plodding evolution of the law didn't stifle new technologies and business models. As copyright law struggled to accommodate digital downloads, developers and the consuming public migrated to the cloud. Rather than downloading purchases to our local hard drives, we accessed music, books, and movies stored remotely, aided by ubiquitous high-speed network connections. Today, a third major shift is underway as subscription streaming services are poised to overtake hard copies, downloads, and the cloud. Services like Netflix and Spotify give subscribers access to massive libraries of material for low monthly fees, prompting droves of viewers and listeners to give up on the idea of buying content altogether.

For consumers, these developments offer obvious benefits. Price, convenience, and selection have improved in the digital era. The widespread embrace of the subscription streaming model, for example, signals a growing demand for low-cost, temporary access to digital media. And since subscriptions offer both flexibility and clarity, they are a welcome addition to the marketplace. But other new ways of acquiring media introduce uncertainty that muddies the waters for people trying to navigate the digital market.

In part, that's because the exhaustion principle and by extension consumer property rights are built around the idea of the tangible copy. But as these new distribution technologies have evolved, the distance between the marketplace as it exists and the marketplace as it is imagined by our copyright laws widens. Each of these shifts in distribution technology has taken us another step away from the copy-centric vision at the heart of copyright law. The failure of the legal system to respond to this disconnect is a major factor in the erosion of consumer property rights, a development that could harm both the public and creators in the long run.

The Hard Copy Era

Since its earliest days, copyright law has evolved, albeit slowly, in response to changes in the ways we copy and share creative works. And for much of copyright history, those changes involved hard copies. This focus on tangible copies influenced the development of the law in a number of ways. Perhaps most important, it enshrined a sharp distinction between the work—the intangible creation of an author—and the copy—the tangible artifact in which the work is recorded. That copy/work distinction is a fundamental assumption of our copyright system. In a world populated with copies, that assumption made sense. But in a market that radically deemphasizes the copy, the utility of the copy/work framework is far less clear.

From its embryonic stages, copyright law focused on tangible copies. With Gutenberg's introduction of the printing press in 1450,[1] control over copying became an imperative for both publishers and governments.[2] Regulations like the Venetian and English printing privileges, which gave exclusive rights to make books to trusted printers, sprung up in the wake of the printing press.[3] Later, the proto-copyrights issued by the Stationers Company—a group of London printers who enjoyed a royal monopoly—also focused on making and selling tangible copies.[4] And the first U.S. Copyright Act in 1790—like its English predecessor, the Statute of Anne—provided exclusive rights to print, publish, and sell books as physical artifacts.

In the twentieth century, copyright law expanded to include not only books, maps, and charts but also dramatic works like plays, musical compositions, visual art, photographs, motion pictures, and later recorded music, architecture, and computer software. At the same time, in addition to making and selling copies, the statutory rights of authors grew to embrace publicly performing and displaying a work, and creating derivative works like sequels and translations.[5] This expansion reflected the technological advances of the day. Live performances of musical and dramatic works had long been economically crucial to some creators. But motion pictures, radio, and television enabled valuable new uses of works that didn't depend on the distribution of physical copies. Copyright holders predictably called for new legal protections to help them profit from those uses.

But throughout this period, the sale of copies remained the core focus of most creative industries, and of copyright law. The fortunes of publishers were tied to the sale of hardcover and paperback books. The music industry enjoyed revenue from radio, but made the lion's share of its profits by selling copies—first sheet music, and later records, tapes, and CDs. And up until the last few years, the software and video game industries were primarily in the business of distributing tangible copies to the public. Even the film industry shifted toward selling home videos, despite its frantic objections to the VCR. Television, because of business models premised on advertising and cable subscriptions, was less concerned with selling tangible copies. But toward the end of the hard copy era, even TV studios profited from DVD and Blu-ray sets. In the early years of the twenty-first century, the copyright industries largely revolved around the economic value of the copy.

This focus on copies is reflected in the fixation requirement. Remember, to be protected at all, a work needs to be fixed in some stable, tangible form. Copyright doesn't protect a poem stored in your memory, but it does once you scribble it on a napkin. Conceptually, the law distinguishes between two forms a work can take—first, the intangible expression in the mind of its creator; and second, a tangible object containing that expression. But the fact that works exist in these two related but distinct forms complicates questions of ownership. As early as 1741, English courts recognized that the copyright in a work was distinct from ownership of any particular copy of it.[6] So the owner of a number of letters written by Alexander Pope, for example, wasn't entitled to publish their contents. Ownership of the physical artifact did not give the would-be publisher the right to copy the underlying work. The U.S. Supreme Court embraced the same principle in 1860 in *Stephens v. Cady*.[7] There the Court held that the owner of a copperplate could not reproduce the map engraved on it. Again, ownership of the work and ownership of the copy were separate questions. Occasionally, courts forgot this lesson and decided that by delivering a physical copy like a book manuscript, an author necessarily transferred their copyright interest.[8] In an effort to underscore the distinction between the work and the copy, Congress provided in the Copyright Act of 1976 that "ownership of a copyright ... is distinct from ownership of any material object in which the work is embodied."[9]

This copy/work distinction has helped resolve disputes over transfers of copyright ownership. But even more important, it has shaped copyright law's exhaustion rules in profound ways. The distinction provided the conceptual framework and vocabulary copyright law uses today to think about the relationship between the rights of consumers and creators. Relying on the copy/work distinction, exhaustion rules have drawn an easily understood line separating those respective rights. Creators own their intangible works; but purchasers own the copies they buy. Of course, putting exhaustion in these terms oversimplifies things a bit. "Ownership" is not a self-defining term. Exactly what rights a copy owner enjoys depends on what rights copyright holders retain in the works. So if Congress insisted that copyright holders get to control public displays of their works even after a sale, owning a copy of a painting would mean something quite different from what it does today. Nonetheless, by articulating the exhaustion principle in terms of copies and works, copyright law takes advantage of our built-in understanding of personal property. Our experiences with tuxedos, cars, and microwave ovens translate reasonably well to the rules surrounding physical books, records, and paintings. By linking consumer rights to tangible objects, copyright law has helped the public embrace exhaustion and accept its limits.

Because our exhaustion rules developed during the era of the hard copy, the way copyright law talks about and conceptualizes consumer property rights is deeply tied to tangible copies. The equilibrium that exhaustion established has worked so well over time because the way works were distributed and sold remained largely unchanged. But an exhaustion principle rooted in the copy/work distinction only makes sense if we are still dealing in things we recognize as copies. Since Gutenberg, copies have been a fact of life. But their place in our digital future is increasingly uncertain.

The Trouble with Downloads

Format changes are nothing new. We used to listen to music on vinyl records, then eight-track and cassette tapes, and most recently CDs. In many ways, the rise of digital downloads looks like just another in a long line of new and improved formats. CD players joined the turntables collecting dust in our collective cultural garages as we marveled over our shiny new iPods. Like earlier format shifts, this one touted many benefits for music fans—increases in portability, convenience, and selection, reductions in price, carbon footprint, and clutter. But digital had downsides too. The browsing experience couldn't compete with a good record store. Digital thumbnail artwork was no replacement for gatefold sleeves or even CD booklets. One factor most of us probably failed to take into account in this trade-off was the impact the move to digital could have on our ownership of the music we buy.

Despite that fact, or perhaps because of it, digital downloads quickly gained market share. Apple launched the iTunes Music Store in 2003. At the time, its catalog was a mere 200,000 songs. Within a decade, iTunes boasted a library of forty-three million tracks collectively downloaded thirty-five billion times, making Apple the largest music retailer in the world.[10] As CD sales dropped and digital sales rocketed upward, paid music downloads surpassed physical media sales. This trend extended to other media, with digital downloads poised to replace hard copies as the primary way we acquire copyrighted material. Paid software and video game downloads rivaled or surpassed brick and mortar sales. And once Amazon released the Kindle, annual ebook sales increased from 10 million units in 2008 to 510 million in 2014.[11] More recently, ebook sales have plateaued, partly in reaction to price increases imposed by publishers.[12] But they remain a major component of the book market.

This shift to digital copies signaled an important shift in the distribution chain for creative works. For physical copies, an author or musician creates a work, often in concert with a large institutional copyright holder. They hand that work off to a manufacturer to produce lots of copies. Records are pressed, books are printed, and video cassettes are manufactured. Those copies get loaded into trucks, shipped around the world, and stocked on retail store shelves. When you buy one of those products, you come home with a new physical artifact containing the work of your choice. Under the digital model, things look quite different. For one, the traditional roles of publishers, labels, and studios are less crucial. Artists can release their own music, authors can self-publish, and independent film makers can find an audience more easily and more reliably than ever before. That's because the digital distribution chain has so successfully reduced barriers to entry. Rather than sending a master copy off to some factory for costly mass reproduction, copyright holders can submit digital files directly to digital retailers. There are no manufacturing costs, no shipping costs, and constraints on shelf space are effectively eliminated. Digital retailers store those files on their servers and make them available to a worldwide audience. When a customer presses the ubiquitous *Buy Now* button, the retailer initiates a transfer of data over the Internet that is then stored on the customer's device.

If that device has a hard drive, the file is stored magnetically on a disc. If it has a solid-state or flash drive, the file is stored electronically in a series of transistors. In either case, the result is the same. A physical object is altered, resulting in a new copy of the work. In this sense, digital distribution isn't all that different from the printing press. Both produce a tangible artifact—a hard disc, flash drive, or printed page—containing the work. And in some ways, copyright law treats stored digital copies the same way it treats more immediately recognizable physical copies. They count for fixation purposes. A novel typed on your laptop is just as fixed as one typed on an antique IBM Selectric typewriter. And they count for infringement purposes. Making unauthorized digital copies of that novel can be just as infringing as an unlicensed print run.

But digital copies differ from earlier physical copies in significant ways. Consider how digital files change the way we transfer copies between people. Imagine you just finished a novel that you are sure your best friend would love. If it's a hardcover or paperback, you simply hand it to them the next time you see each other. If they live across the country, maybe you mail it instead. What if instead, you bought an ebook? Assuming your ebook isn't one of the few titles that qualify for Amazon's licensed digital "lending" program, how do you let your friend borrow your copy? That depends on what you mean by "your copy." If the copy is the physical embodiment stored in your Kindle's memory, you could lend them your device. Of course, that means handing over your entire digital library and an expensive piece of hardware. It would be like if lending a friend one hardcover meant giving up the entire contents of your bookshelf. And if your device also contains your personal email, documents, and photos, lending out your copy is even more problematic.

The other, more reasonable option is to keep your device and just send your friend the file. You could email it, save it to a cheap flash drive, share it via Dropbox, or use one of the dozens of other ways we move data between people. The problem is that each of those methods of sharing your ebook requires making one or more new copies of the file. And that's precisely what copyright law would seem to prohibit. The first sale doctrine gives owners the right to transfer their copies, to pass one object from person to person, but it remains unclear how courts will view the creation of new copies to facilitate transfers.

The same is true if you want to resell your digital purchases. The orthodox understanding of copyright law says that making a new copy, even as part of the process of transferring a file to a new owner, is copyright infringement. ReDigi, a company that launched the first online resale marketplace for digital music late in 2011, found that out the hard way. Assume—hypothetically of course—that you bought a copy of "The Sign" by Swedish pop group Ace of Base from the iTunes Music Store in a fit of 1990s nostalgia. And let's say you later regretted that impulse purchase. ReDigi allows you to resell that track to another equally nostalgic buyer and recover some of the 99 cents you paid for it in much the same way you can still sell used CDs and records.

ReDigi, well aware of the legal risks, designed its system to ensure that only one copy of the file existed at any particular time. So you decide to unload your copy of "The Sign." That copy lives on your laptop's hard disc, encoded in the language of magnetic charges. To sell your copy, the file has to be uploaded to ReDigi's server, where it waits for its lucky new owner. If ReDigi's software simply uploaded the file, two copies would exist—one on your hard drive and another on the ReDigi server. ReDigi wanted to recreate as closely as possible the mechanics of a traditional used sale, one where a single copy moves from one owner to another. To achieve that, as each of the thousands of packets of data that make up the file were sent over the Internet, ReDigi deleted that data from your hard drive. So your local copy disintegrated piece by piece as it was being reconstructed on the ReDigi server. That way, two complete copies never existed at any one time. ReDigi argued its process simply migrated a copy from point A to point B, just like mailing your hardcover book or taking an unwanted CD to the used record store. As in a traditional sale, the seller starts out with one copy and ends up with none. And the buyer starts out with no copies and ends up with one.

Unlike a traditional used record store, ReDigi took great effort to make sure the seller wasn't keeping an extra copy for themself. But neither copyright holders nor the courts saw it that way. Capitol Records sued ReDigi for reproducing copies of its tracks.[13] The question for the court was whether ReDigi's software succeeded in moving the file from one location to another, or if it simply made a new unauthorized copy. The court sided with Capitol. Because the process resulted in the work being encoded on a new material object—ReDigi's server—a new copy was made. And that was true even if ReDigi destroyed the original as part of the upload process. The lesson from *ReDigi* is that even if you delete your copy after transferring a file, you've likely violated copyright law by reproducing the work.[14] If that's true, the application of our legal rules to digital copies is inconsistent with the expectations about lending and reselling developed in the hard copy era.

This isn't the first time copyright law has encountered this sort of mismatch. Congress recognized the same problem when it expanded copyright to protect computer software. Then, end users faced similar problems.

They couldn't run their software, create backups, resell, or transfer their purchases without potentially running afoul of the law. As Congress recognized, that outcome would be at odds with the idea that users owned the software they bought. So it enacted a new provision, section 117 of the Copyright Act, to address the situation.[15] That section guarantees users the rights to reproduce copies of software they own for preservation purposes, to adapt them to run in new software or hardware environments, and to transfer copies of the software they purchase so long as they delete the copies in their possession. So users could, for example, resell their software even if it meant making additional copies in the process.

Section 117 was an explicit attempt to extend the longstanding commitment to exhaustion to digital copies. But as we explain in chapter 4, its practical effectiveness has been undermined by license agreements that redefine software sales. Equally important, Congress has never extended the kinds of rights recognized in section 117 to other forms of media, even as digital downloads displaced hard copies. But even if it had, the digital download looks more and more like a transitional technology.

The Cloud of Uncertainty

Even if you don't understand exactly how the "cloud" works, you're probably familiar with the term. Cloud computing allows users to remotely access resources like data, programs, processing power, and storage from a variety of devices. Rather than keeping all of your software and data on your local desktop or laptop, you can use phones, tablets, or other network-connected devices to access files or run programs stored on remote servers. When it comes to the distribution of copyrighted content, the cloud allows companies to sell you music, movies, and books without requiring you to download them. Your files are stored in your Apple iCloud account or Amazon Cloud Locker. When you want to hear a song or watch a video, they can be streamed to your device anywhere you have a data connection.

Like new forms of distribution that came before, the cloud was driven by the technology of the day. Digital downloads made sense when people wanted dedicated media playback devices with large storage capacities. They would download music and movies to their home computers, which functioned as central hubs for their devices, and sync those files to their media players. You could buy a 160GB iPod and cram your entire digital library onto it.

But iPods soon gave way to smartphones, tablets, and other multipurpose, mobile computing devices. These devices jettisoned cheap, high-storage capacity hard drives in favor of more expensive, lower-capacity flash memory to conserve space, weight, and battery life. So there was no longer enough room to lug around your entire media collection on a device's internal memory. Even if there had been, people were tiring of the hassle of syncing devices through their home computers. The cloud took advantage of increasingly ubiquitous, reasonably affordable, high-speed mobile data networks to solve the storage and syncing problem. All of your digital stuff could be available all the time through the wonders of the cloud, without the need to download a single copy.

The cloud is not without its drawbacks, however, most of which are byproducts of the lack of physical possession implicit in cloud-based distribution. In the hard copy and digital download eras, you literally possessed your copies. They were stored on shelves, in attics, and on hard discs. But if you possess copies in the cloud, you do so only virtually. They aren't actually on your device; that's the whole point. For one, the cloud creates some real concerns about privacy. Since a digital record is created every time you access a file, your reading, listening, and viewing habits are being closely tracked.

The lack of physical possession also means your ability to access your purchases depends on the cloud service provider keeping up their end of the bargain. The provider might suffer an outage, or the title you purchased—like *1984*—might be pulled from the service. Apple's iTunes terms specifically address this possibility: "Apple and its licensors reserve the right to change, suspend, remove, or disable access to any iTunes Products, content, or other materials comprising a part of the iTunes Service at any time without notice. In no event will Apple be liable for making these changes."[16]

So if you buy a movie or album and store it on Apple's cloud server, your purchase can disappear if Apple or the copyright holder decides for any reason to remove it. Instead, retailers might simply stop supporting their cloud offerings if they are no longer profitable, or they might go under altogether. A future without Apple and Amazon is hard to imagine, but so was one without Lehman Brothers, Enron, and Woolworth's. And the fortunes of even the most highly valued technology companies are volatile. Just ask Yahoo, Myspace, or Apple circa 1997.

So what does the cloud mean for consumer property rights? Without physical possession, consumers can't be entirely confident in their ability to access their purchases in the future. Their rights to lend, resell, or otherwise transfer those purchases are even more uncertain. Exhaustion, as we've seen, traditionally has been premised on ownership of a copy. But the cloud, it turns out, doesn't result in a single identifiable copy. Instead, it creates a tangled web of potential copies. It's not clear who owns them or whether they even count as copies for copyright purposes.

At the risk of oversimplification, we can think about two distinct sets of copies in a cloud system. Some copies are stored on the cloud server, and some copies are stored—potentially—on the user's device. Let's start with the server copies. They are stored long-term and are certainly fixed for copyright purposes. But who owns them? Before turning to how a court might answer that question, let's consider three analogies to more familiar examples.

First, we could think of the cloud server copy like a film reel at your local movie theater. The reel is owned and possessed by the theater.[17] You pay to see the movie, but never possess, much less own, the reel. With the cloud, you pay to listen to a song or watch a program, but the cloud provider owns the copy—literally, the hard drive in its server. Any rights you have would be rooted in contract, not property. Second, maybe the cloud copy is more like a library book. The server, like the library, is full of copies of works. When you want to access your purchase, the work is plucked from the shelves and transmitted to your device in much the same way you can select a library book and take it home. But again, you don't own the copy on the cloud server any more than you own a book from the library. Third, perhaps the cloud copy is more like a family heirloom in a safe deposit box. You pay for a movie, and it waits for you on the cloud server until you are ready to access it. Just like grandpa's stamp collection at the local bank, you own it even if you don't currently possess it.

Copyright law can easily make sense of these first two analogies because they fit into the copy/work dichotomy. For the film reel, control over a tangible copy regulates access to the intangible work shown on the screen. For the library book, a tangible copy changes hands, but not permanently. But from the perspective of the purchaser of cloud content, both of these analogies are unsatisfying. Surely, you might think, "buying" *Upstream Color* for $12.99 gets you something more durable for your money. But identifying what that something is—in terms copyright law can comprehend—is a challenge. Its vocabulary is limited to tangible copies and intangible works. Unlike your personal property interest in Grandpa's stamps, you don't literally own part of Apple's server. Your property right relates to something less concrete. But at the same time, it isn't the intellectual property interest in the movie. A $12.99 purchase doesn't make you the copyright holder in *Upstream Color*. Conceptually, it makes much more sense to talk about an intangible property right in the file—the collection of bits that encodes the movie—detached from any particular physical copy. In much the same way you can own and transfer stock—an intangible interest in a corporation—you can own and transfer rights in your cloud purchase.

From a technical perspective, those are rights the cloud provider could easily accommodate. Let's say you wanted to lend the movie you bought to a friend. Amazon, for example, could easily transfer rights to the file by associating it with your friend's user account rather than yours. When your friend logs in to their account, the file would be there for them to access. But under your account, the file would be disabled. This is how Amazon's existing Kindle ebook "lending" program works.

Of course, the precise scope of consumer intangible property rights would be determined by the rules of exhaustion, just as our personal property rights are today. And the division of rights between creators and consumers might look different for intangible property, but the key is that your rights would be determined by default property rules, not the minutiae of a EULA. By calling it property, the law would shift the balance of power from sellers to buyers and responsibility for defining our rights from lawyers at Amazon and Apple to courts and legislators.

To be clear, U.S. law hasn't yet recognized intangible property interests in digital media. So the question of how courts today would think about cloud copies remains. Courts have adopted two very different ways of evaluating new technologies that challenge embedded assumptions of copyright law. One approach—familiar from *ReDigi*—closely examines the design and operation of a technology. There the court focused on maintaining a careful ledger of copies, rather than evaluating those technologies from the perspective of the end user.

In a recent case called *ABC v. Aereo*, the Supreme Court took the opposite tack. Aereo offered its subscribers access to broadcast television programming over the Internet by constructing an elaborate system of thousands of dime-sized antennae, each assigned to an individual subscriber. When a subscriber chose a program, their antenna would tune to the appropriate station, and the show would be recorded by a server to hard drive space dedicated to that subscriber. Aereo's system was designed with the law in mind. By making sure each antenna and each recording corresponded to a single subscriber, it hoped to design around copyright's public performance right. In holding Aereo liable for infringement, the Court emphasized the "viewing experience of Aereo's subscribers" and discounted the importance of "behind-the-scenes" details about the operation of the technology.[18]

Both of these approaches have merit, and we don't mean to suggest that either is inappropriate. Developers shouldn't be penalized for designing systems that comply with the letter of copyright law. But when it comes to cloud copies, there are two reasons we think it makes more sense to focus on end user experience rather than technical design choices. First, we have been trained to ignore what happens under the hood. The engineers behind cloud services have done a remarkable job of shielding us from their complexity. Those services are intuitive. True to Apple's philosophy, they just work. The downside of such high usability is that it obfuscates details about precisely how these services operate.

Second, an approach that emphasizes technical details misunderstands what is valuable about the cloud. Back when individual copies were valuable, long-lasting artifacts, keeping a running ledger of copies made sense. But this preoccupation with counting copies is outdated. Copies today are cheap, disposable things. We are awash in a sea of copies that flit into and out of existence all the time. They are created, used, and discarded constantly. What matters to consumers are reliable rights to access and use a work. And those rights, as property theory makes clear, don't have to be tied to any particular tangible object. But until copyright law rethinks the central role of the copy, ownership of cloud purchases will remain a challenging question with no obvious answer.

What about the copies on your own device? There the tough question isn't so much about ownership, but whether we have a copy at all. Here we need to distinguish between downloading and streaming content. If a cloud customer saves a file to their device—one that they can access during a long flight without Internet access, for example—that looks like a standard download. There's a stable, lasting copy stored to the memory of their phone or tablet. Streaming, in contrast, allows the customer to listen to music or watch a video without permanently saving a file to their device. It isn't intended to result in a lasting copy.

Nonetheless, some courts have held that data stored even temporarily in the random access memory (RAM) of a device can count as a copy for copyright purposes. If so, using a digital file—reading a book or playing a song—means creating new copies. When you open a file on your laptop or your mobile phone, your device is accessing data in long-term storage, on a hard disc or flash drive, and recreating it in its RAM, the short-term storage used to display and manipulate data. But the rules for precisely how long such data can be in RAM before a fixed copy is created are far from clear. So despite the emphasis copyright law places on keeping track of copies, it has a surprisingly difficult time figuring out whether a copy even exists.

Although this problem has become more pronounced in recent years, it turns out that it is hardly a new challenge. Copyright law, in fact, has been struggling to answer that question for more than a century. When the player piano hit the market in the late nineteenth century, composers and music publishers were in the business of selling sheet music that people took home to play on their pianos. But by combining a pneumatic mechanism and perforated paper rolls, player pianos enabled people to listen to music at home without a musician on the premises. Music publishers argued that piano rolls were infringing copies of their compositions. But after years of litigation, the Supreme Court in *White-Smith v. Apollo* unanimously rejected that argument.[19] According to the Court, piano rolls weren't copies at all since no one, including the makers of piano rolls, could look at a series of tiny perforations and discern the music it contained. Copies, the Court explained, are limited to those forms in which a work can be seen, read, or understood by the human eye. Today we have a much broader notion of the copy, but *White-Smith* shows how new technology can frustrate efforts to apply laws written for an earlier era. That's just as true today as it was a century ago. In fact, the cloud has given rise to its own existential crisis over copies.

Cablevision is a large cable television provider. In 2006 it launched a cloud-based Remote Storage Digital Video Recorder (RS-DVR) for its subscribers. Most DVRs come equipped with a large hard drive to store recorded programs. Cablevision's product stored recordings made by subscribers on remote servers in a central data center instead. In that data center, Cablevision used a device called the Broadband Multimedia Router (BMR) to send the constant stream of video for each cable channel to the servers that stored recorded programs. As it did so, the BMR briefly loaded the video into temporary memory buffers for a period of a second or so.

Cartoon Network sued Cablevision for copyright infringement, alleging that these buffers created infringing copies of its television programs.[20] The case turned on whether or not the programs were stored long enough to count as copies. How long must a work be stored before it counts as fixed? One influential early case, *MAI v. Peak,* held that Peak created copies when it turned on MAI's computers and loaded MAI programs into memory. According to the *MAI* court, if information stored in the memory of a computer could be perceived or reproduced, it was fixed regardless of how long it was stored.[21] If that were true, Cablevision made copies in its buffers. But the *Cablevision* court disagreed. It held that the data must last for more than a "transitory duration" before it counts as a copy. The court was convinced that 1.2 seconds wasn't long enough to create a copy, but beyond that, it didn't offer much guidance. So when you use a cloud service to stream a movie or song to your device, copyright law has no clear answer as to whether you even possess a copy.

The *Cartoon Network* case, much like *ReDigi*, demonstrates how copyright law struggles to consistently and clearly identify copies in the digital environment. That fact puts consumer property rights at risk so long as exhaustion rules are tied to ownership of a copy. Without copies, under current law, there's simply nothing to own. The next major shift in distribution, however, suggests that ownership isn't important to everyone.

Crossing the Stream

All of this talk about ownership assumes that we will still be buying music, movies, and books in the near future. If current trends hold, however, à la carte purchases could soon be the exception rather than the rule. Every day, more people are choosing digital subscription services over individual purchases. Although we call them subscriptions, these services don't have much in common with analog magazine or newspaper subscriptions. If you decide not to renew your *National Geographic* subscription after a year or a decade, you still own the stack of issues they've sent you. If you cancel your Spotify subscription, you don't keep anything. Instead, the digital subscription model allows you to pay a flat monthly rate—or patiently endure advertisements—in exchange for access to large libraries of streaming content. And for many of us, that's an attractive proposition.

Netflix and Hulu led the way, launching online video services in 2007. Since then, Netflix has become one of the most popular content providers on the Internet. The service boasts roughly sixty-nine million subscribers and accounts for as much of a third of all Internet traffic.[22] In 2014, its revenue exceeded $5.5 billion. On the music side of things, Spotify claims seventy-five million active users, about twenty million of whom are paying subscribers.[23] It recently broke the billion-dollar revenue barrier for the first time.[24] Not surprisingly, this subscription model is being applied to other forms of content as well. In 2014, Amazon launched Kindle Unlimited, which gives subscribers access to a growing ebook library. Meanwhile, services like PlayStation Now, EA Access, and Gametap offer subscriptions for online video game libraries.

Rapid gains in market share by these services point toward a future in which subscriptions, not purchases, will be the primary way we access copyrighted works. By 2016, revenue from digital distribution of movies—including subscriptions and purchases—will eclipse physical sales.[25] The bulk of that money will come from subscriptions. Even today, subscriptions account for nearly three times as much revenue as digital downloads.

A similar story is playing out in the music industry. In 2014, as CD sales continued to plummet and paid downloads dropped by roughly 10 percent, streaming music services like Spotify grew by a staggering 54 percent as users streamed 164 billion songs.[26] By 2018, streaming services are projected to account for nearly 40 percent of music industry revenue. Already in Europe, Spotify's revenues are overtaking Apple's music download figures. Given these trends, Apple spent $3.2 billion to acquire Beats Electronics, driven at least as much by its interest in the successful Beats Music subscription service as it was by the company's better-known headphones. Apple launched its own subscription streaming service in 2015.[27]

The public seems to be sold as well, and for good reason. Subscription services make a compelling case in terms of price, selection, and flexibility. All-you-can-eat subscriptions for Netflix, Spotify, and Kindle Unlimited cost less than $10 per month. For that price, you might be able to buy a single ebook, digital album, or movie. Instead, subscription services offer unlimited access to massive collections of works. Spotify boasts thirty million tracks; the Netflix streaming library tops out at over sixty thousand movie and television titles; and the Kindle Unlimited collection includes over a million books.[28] These services don't include every new blockbuster or bestseller, and music fans enjoy access to a much more complete library than movie buffs or bibliophiles. Nonetheless, users seem generally satisfied with both the quantity and quality of options.

Netflix has used its massive success as a springboard to becoming a leading content producer, with exclusive programming like *House of Cards, Master of None*, and the resurrected *Wet Hot American Summer*. Others like Amazon and Hulu are pursuing a similar strategy with varying degrees of success. Another key selling point for subscription services is their compatibility with nearly the full range of media devices. You can stream Netflix to your laptop, tablet, smartphone, television, or game console. The same is true for Spotify and most other competitors in this space. That allows users the degree of portability the cloud helped teach them to expect.

You might look at the basic business model of the subscription streaming service and wonder how different it is from familiar twentieth-century approaches to distribution. Consumers pay, either by ponying up a monthly fee or by sitting through advertisements, in exchange for the ability to enjoy a curated collection of programming. That sounds like a reasonably accurate description of broadcast or cable TV, or even terrestrial radio. So what sets services like Netflix and Spotify apart? And what explains their massive explosion in popularity in recent years?

In part, the answer is control. Radio and television have always been fundamentally passive media. You sit back and hope the DJ plays your favorite song. Television required viewers to wait until their program of choice aired each week. But streaming services allow users to browse their libraries and watch whatever movie or hear whatever song they want. Right now. And if you want to watch thirteen hours of *House of Cards* without leaving your couch, Netflix is more than happy to oblige you. That degree of choice and immediacy distinguishes subscription services from cable and broadcast. It also makes those services a much closer substitute for purchases, and at a lower price point.

Interest in these services is easy enough to explain, but service providers and content producers have reasons to favor the subscription model aside from simply satisfying consumer demand. Strategically, it offers a number of benefits. Compared to sales-based models that wax and wane depending on a host of factors, subscriptions generate relatively predictable and reliable revenue streams. They also yield mountains of valuable data about subscribers, their viewing habits, and preferences that can be used to tailor the service and produce new programming, as Netflix did when it ordered a full season of *House of Cards* without the once-obligatory pilot episode. For some subscription video providers like HBO, ESPN, and Nickelodeon, launching a standalone digital subscription service allows them to reduce their reliance on the cable company to play the role of middleman. It also provides them an avenue for reaching the increasing number of cord cutters without cable subscriptions. Streaming services also allow movie and television studios to bundle large libraries of old and relatively low-value content with some new, high-value programming. By doing so, they can squeeze additional revenue out of movies and shows that would otherwise be collecting dust in a vault.[29]

Subscription services are also an effective strategy for reducing the effects of widespread copyright infringement on the Internet. By setting the price point so low, Netflix and Spotify can attract subscribers who might otherwise get their movies and music from the Pirate Bay. More fundamentally, by shifting from selling an easily copied product to selling a hard-to-copy service, Netflix and its cohort trade on the value of convenience, curation, and recommendation. By doing so, they insulate themselves from the harsh reality of the Internet—that copies are free for the taking.

Finally, by moving away from the sale of copies, producers get the added benefit of reducing resale. In the era of physical media, used copies competed with new ones, reducing sales and driving down prices. Since Netflix doesn't distribute copies, physical or digital, secondary markets have no chance to develop. There's no small amount irony in this fact. Netflix originally rose to prominence as a DVD-by-mail company. In that line of business, the first sale doctrine and resale markets were crucial to its success. Although Netflix bought the majority of its DVDs in bulk directly from movie studios at discounted rates, those negotiations took place against a backdrop of widespread availability of DVDs on the open market and the legal right to lend them. And in at least one instance, after the Weinstein Company signed an exclusive distribution agreement with Blockbuster, Netflix resorted to buying DVDs at retail to meet subscriber demand.[30]

The impact of subscription services on individual creators is much trickier to untangle. First, there's the question of revenue. Do subscription services put money in the pockets of creators? It's too early to tell whether or not Amazon's experiment with Kindle Unlimited will be a boon for authors. Some report significant boosts in readership and revenue, while others allege that their sales are shrinking since the service's launch. For movie makers, the answer is more certain. Netflix was once seen as an extra unforeseen revenue source by Hollywood, but today streaming revenue is part of the calculus that makes or breaks a potential project. Films and TV shows get produced based in part on their likely value in the subscription market. There has certainly been plenty of squabbling over how much Netflix should pay for streaming rights, and sometimes titles get pulled, occasionally by the thousand. But there's been no massive outcry against subscription services from studios, producers, or directors.

The same can't be said for music. From well-established performers like Thom Yorke, David Byrne, and Beck to lesser-known artists like Jason Isbell and Phil Elverum, musicians have voiced concerns about the paltry sums they say performers and songwriters receive from streaming services. Spotify pays just fractions of a penny each time a song is streamed, an amount many find not only insufficient, but insulting. In part, the size of these streaming royalty checks reflects a simple economic reality: people are not willing to pay as much for a product that is less valuable. Recording artists make significantly more money from CD sales because they give owners something of enduring value. If you own a CD you can play it as many times as you want; you can lend it to a friend; you can resell it. Despite what Garth Brooks may have thought, when we eliminate ownership in favor of temporary access, we get Spotify—not some artist-friendly utopia.

Spotify counters its critics by noting that all those microroyalties add up. The service has paid out over $2 billion—70 percent of its revenue—to copyright holders for the rights to its streaming catalog.[31] That's not to say Spotify couldn't pay more, at least in theory. If the market would bear it, they could increase subscription fees or their advertising rates. Or perhaps they should hand over an even higher percentage of their revenue. But those efforts wouldn't solve the problem. Ultimately, Spotify has only so much control over how much money makes its way into the pockets of artists. Those payments are filtered through record labels, music publishers, and collecting societies, each of which takes its own cut. It turns out that most of the $2 billion paid by Spotify has replenished the coffers of record labels, while very little has gone to artists. But that fact is a function of the contracts between recording artists and their labels, not the real or perceived stinginess of streaming services.

Of course, no one is forcing copyright holders to license their music to Spotify. If they don't like the deal being offered, they can refuse it. And many artists have, including AC/DC, The Beatles, Garth Brooks, Led Zeppelin, and Radiohead. But no opt-out was met with the Internet-wide hue and cry heard when Taylor Swift broke up with Spotify in 2014. After her record *1989* sold nearly 1.3 million copies in its first week, the strongest debut in over a decade, Swift decided to pull her catalog from Spotify. Many people slammed her decision as a rich pop star's attempt to boost her already singularly strong record sales by cutting off free access to her songs. Swift, it was argued, was simply pursuing her own short-term economic interests—interests that, given her position in the music industry, were poorly aligned with all but the tiniest circle of ultra-popular recording artists.

No doubt, the economics of record sales motivated Taylor Swift. But there's good reason to think her decision was about something more than maximizing sales of her current record, one that needed little help in that regard. More than most musicians with her level of success, Swift seems interested in building—and publicizing—a close connection with her listeners. She dances with them in her music videos, buys them lunch, comments on their Instagram photos, sends them Christmas gifts, and shows up for their bridal showers. Taylor Swift doesn't want to rack up plays, she wants to cultivate fans.

Casual listeners might play her current hit for free on Spotify, and they might even sing along. But loyal fans will not only buy Taylor Swift records, they will shell out for concert tickets, t-shirts, and all manner of This Sick Beat® merchandise.[32] They will establish connections that span a career, or even a lifetime. That's the level of commitment Taylor Swift wants from her fans. Months before the release of *1989* and her self-imposed Spotify exile, Swift penned an editorial in the Wall Street Journal. She wrote in part: "People are still buying albums, but now they're buying just a few of them. ... The way I see it, fans view music the way they view their relationships.

Some music is just for fun, a passing fling. ... However, some artists will be like finding 'the one.' We will cherish every album they put out until they retire and we will play their music for our children and grandchildren. As an artist, this is the dream bond we hope to establish with our fans."[33]

Here's where ownership comes back into the picture. Listeners who choose to spend ten dollars on a particular record by a particular artist, rather than on a subscription to an undifferentiated mass of content, are more likely to feel invested in those artists. So meaningful personal property rights could benefit not only consumers, but creators as well. If we own a Taylor Swift record—as opposed to merely listening to it on the radio or streaming it online—it means more to us. Because the things we own can help define who we are, buying *1989* identifies you, both to others and to yourself, as a Taylor Swift fan in a way that a Spotify playlist might not. That, in turn, helps transform casual listeners into the sort of fans who can sustain an artist's career.

The value we place on ownership also finds support from the field of behavioral economics. Over the past twenty-five years, dozens of experiments have established what researchers call the endowment effect—the widespread tendency of people to assign greater value to things they own. In one well-known example, researchers gave some participants coffee mugs. When presented the opportunity to sell or trade their mugs to other participants, mug owners demanded nearly twice as much compensation as nonowners were willing to pay.[34] Subjectively, they valued the mugs they owned well above the market rate.

What explains these vastly different assessments of the value of an otherwise ordinary mug? Some have suggested that the endowment effect is the result of loss aversion—the idea that people are more motivated by the fear or regret associated with loss of an item than the enjoyment of gaining it. But more recent research shows that we place greater value on the things we own *because* we own them.[35] The association between an item and its owner means that we value things we own far more than things we simply use. And as that sense of ownership grows stronger, so does the value we place on the item. Recent research has also shown that the endowment effect is no less pronounced for digital goods than it is for physical ones.[36] So if a Taylor Swift fan owns her *1989* mp3s, we should expect her to value them in much the same way owners of *1989* CDs or vinyl do.

The psychological value of ownership might also suggest one reason for the flagging sales of digital downloads. We are used to getting reliable property rights in exchange for the money we spend on music. In the past, if we wanted fleeting access, we'd listen to the radio for free. But when we spent money on music, we got something lasting and transferable. As we detail in chapter 5, many consumers misunderstand precisely what rights their digital download dollars are buying them. But as more people understand the limited value that downloads offer, we shouldn't be surprised to see steeper decreases in digital sales revenue. If digital sales were sales in the true sense of the word—if they were transactions that gave users property rights—we might see very different consumer behavior.

In fact, some of us are still willing to pay a premium for property rights. The only format that can rival the growth rate of streaming subscriptions in recent years is vinyl. In 2014, vinyl record sales increased more than 50 percent over the prior year.[37] In absolute terms, the number was a relatively modest 9.2 million units, but it was the largest vinyl tally in decades. That upward trend continued in 2015.[38] Even though vinyl is generally the most expensive way to get new music, there are plenty of reasons to prefer it. Aside from the appeal of higher fidelity and better packaging, when you buy a record you are bargaining for the full range of property interests associated with a purchase, rights that are not contingent on license terms, digital permissions, or even an Internet connection.

The rise in these two very different approaches to consuming music—subscription services and vinyl records—highlights the importance of consumer choice. Not everyone wants to rent their music, and not everyone wants to own it either. These choices aren't right or wrong. They are preferences that vary between individuals and over time for a host of reasons. Luckily for most works today, we have options. But both consumer behavior and industry strategy are limiting the choices available to those of us who prefer property to conditional access. Bookstores and record stores across the country, both big and small, have shuttered. For many of us, that means the immediacy of an in-person retail purchase has been replaced by online ordering, shipping costs, and days of waiting. That makes it hard for analog copies to compete with instant access to digital content, especially when the two formats are offered on the same Amazon product page.

Equally troubling, some content is available exclusively in one format or from one service. For years, a handful of big-name recording artists refused to sell music through iTunes. Others made music you could only buy from Apple. Amazon boasts over half a million titles exclusively available on its Kindle Store. Many works that were once available in a variety of physical formats are moving to digital-only distribution. Fox recently announced, for example, that it would no longer sell new seasons of *The Simpsons* on DVD or Blu-ray in favor of streaming delivery.[39] When works are available only as digital downloads, it limits consumer choice. For libraries, it can interfere with their core function. Apple and Amazon licenses prohibit lending, and they won't—or more accurately, can't—negotiate individual terms for libraries and educational institutions committed to preservation and patron access. Since those works aren't available in a property-friendly format, they are effectively excluded from library collections.[40]

Other works are available only through subscription. You can't buy the digital version of the *Compact Oxford English Dictionary*; you can only access it through a monthly subscription that requires an Internet connection.[41] Adobe's latest creative applications like Photoshop and Illustrator are now available exclusively through the company's Creative Cloud, a monthly subscription service.[42] Unless it reverses course, Adobe will never sell a new copy of Photoshop, effectively suffocating the used market. The shift to digital distribution is so pronounced that Microsoft made headlines when it made the decision to release Windows 10 in a decidedly retro format: on a disc, in a box, sold at brick-and-mortar retail locations.[43]

The copy, at least for the time being, is out of fashion. But as a legal concept, the copy remains as important as ever. Even as copies escape our possession and disappear from our experience, copyright law continues to insist that without them, we only have the rights copyright holders are kind enough to grant us. As we discuss in chapter 4, those rights are often impermanent, nontransferable, and conditioned on ongoing permission. In short, they are not property rights.

4 Ownership and the Fine Print

Imagine you walk into your local haberdashery in search of some new hats. You try on a few bowlers, some deerstalkers, a pork pie hat, even a fez or two. After settling on a favorite, you take note of the price tag—$100—and make your way to the counter to pay. On your way to the cashier, you notice some language printed inside the brim. It reads:

THIS HAT IS LICENSED, NOT SOLD. BY PAYING THE ASKING PRICE, YOU ARE ENTITLED TO WEAR THIS HAT AS OFTEN AS YOU LIKE. YOU MAY RETAIN POSSESSION OF THE HAT INDEFINITELY, BUT YOU ARE NOT PERMITTED TO RESELL, LEND, OR OTHERWISE TRANSFER IT WITHOUT THE EXPRESS PERMISSION OF THE MANUFACTURER.

What should we make of this language? There are at least three ways we might think about the legal impact of this sort of notice. First, it might mean, as the hat maker probably intends, that you don't own the hat after forking over your $100. The notice could transform a deal that looks like a sale—a transfer of ownership from one party to another—into something less—mere permission to possess and use the hat. Second, the notice might form the basis of a contract. By buying the hat, you would become its owner, but you would also be promising not to transfer it. You would have the power to resell it as a matter of property law, but you might have to pay damages under the contract if your transfer causes provable harm to the manufacturer. Third, the notice might be utterly ineffectual as a legal document. It could still confuse or intimidate some hat owners, dissuading them from loaning or reselling their chapeaus. But as a legal matter, it wouldn't impose any obligations or restrict your behavior.

Regardless of which of these interpretations is right, most of us would find this sort of language surprising. Puzzled, we might ask the sales clerk for an explanation. Or maybe we would simply refuse to buy a hat so burdened by unreasonable demands. What many of us fail to notice, however, is that we encounter this sort of language on a nearly daily basis. It is attached to both digital and tangible goods that we assume, rightly or wrongly, we are buying. The boxed software you purchase, the apps and games you download, the digital books, music, and movies you buy, your smartphone, and even your car all come bundled with similar restrictions. But this language is typically buried within the thousands of words of terms and conditions, liability waivers, warranty information, and prohibitions on the development of nuclear weapons—no, really[1]—that make up End User License Agreements, where it blends into the informational white noise most of us are skilled at ignoring.

In the simplest terms, a license is a grant of permission to engage in some behavior that would otherwise be prohibited. You need a license to drive a car, a radio station needs a license to broadcast over the public airwaves, and James Bond needs a license to kill, all for the same reason. Without permission, those activities are against the law. Sometimes that permission comes from the government, and other times from private parties. If you enter your neighbor's property without a license, you are a trespasser; with a license, you are an invited guest.

But modern license agreements have evolved into something else altogether. They create private regulatory schemes that impose all manner of obligations and restrictions, often without meaningful notice, much less assent. And in the process, licenses effectively rewrite the balance between creators and the public that our IP laws are meant to maintain. They are an effort to redefine sales, which transfer ownership to the buyer, as something more like conditional grants of access.

A troubling number of courts have embraced these efforts. They have decided that products remain the property of IP rights holders even after an apparent sale, so long as the license recites the proper incantations. You may have paid for the CD-ROM containing your favorite software program, but according to this reasoning, that plastic disc belongs to the software maker, not you. And your use of it is constrained not by the public law of copyright, but the private law of the license. This way of thinking has its strongest foothold in the world of computer software, but it is unlikely to remain confined to that corner of the economy. The licensing mentality has already spread to digital media and, just as problematically, to tangible goods.

By allowing license terms to redefine transactions and strip consumers of ownership, courts are taking power away from the public lawmaking process and vesting it in the hands of private IP rights holders. Licenses function as a form of privately made law that allows rights holders to modify, supplement, and contravene IP law at the expense of the customers who pay for their products.

The Fine Print

So what do licenses actually say? Most of us have no idea, and for good reason. License agreements are long, inscrutable, and full of bad news. They are the Lars von Trier films of legal documents. The form and substance of license agreements discourage consumers from reading them, which perversely reinforces their worst attributes.

Let's start with their length. The current iTunes Terms and Conditions are over 19,000 words, translating into fifty-six pages of fine print, longer than *Macbeth*.[2] Not to be outdone, PayPal's terms weigh in at 36,000 words, besting *Hamlet* by a wide margin. The demands of these prolix legal documents are jaw-dropping. Take Adobe's Flash, a software platform installed on millions of computers each day. Assume the average user can read the 3,500-word Flash license in ten minutes—a generous assumption given the dense legalese in which it is written. If everyone who installed Flash in a single day read the license, it would require collectively over 1,500 years of human attention. That's true every single day, for just one software product.[3] Imagine what would happen if you tried to read every license you encountered.[4]

Regardless of their length, license agreements are hard to comprehend. They are documents drafted by lawyers, and their primary function is to define legal rights and limit liability, not to communicate clearly and effectively. As a result they are overflowing with defined terms, technical jargon, unnatural turns of phrase, and complex sentence structure. Unlike the accessible and simple marketing language used to sell products, the legal language that defines those transactions—or claims to at least—often requires a postgraduate education to understand.[5]

The length and complexity of license agreements mean that they impose significant costs on the public. Reading and understanding a license requires lots of time and mental energy. For most people, simply wading through the terms won't be enough to actually understand the license. You might need to do some independent research or consult a lawyer—a suggestion some licenses make, presumably with a straight face. But often the cost of reading the license outweighs the value of the product. Who in their right mind would read a 19,000-word license before making a 99-cent purchase from iTunes?

Not surprisingly, the vast majority of us simply throw up our hands and ignore licenses altogether. One recent study shows that as few as one out of a thousand software shoppers even glanced at the text of license agreements. And most who did spent only a few seconds perusing the terms.[6] Even Chief Justice John Roberts, hardly known for his casual disregard for legal obligations, can't be bothered to read EULAs.[7] It hardly seems fair to expect more from the average person.

License terms are not negotiable. So there's little to gain from a careful reading. Suppose you carefully examine the Flash license and find some objectionable term. Perhaps it limits you to a single installation of the program or disclaims liability for damage to your computer. What exactly are you going to do about it? Adobe is not going to negotiate a new license with you. They won't even entertain the idea. So your choice is simple. Either use the product—and live with the license—or don't. Take it or leave it.

Intentionally or not, rights holders and retailers have managed to nearly universally dissuade their customers from reading the terms that purportedly govern their purchases. And if the public rationally avoids investigating licenses, there is little marketplace incentive to offer more consumer-friendly terms. Better terms would simply go unnoticed. When software maker PC Pitstop included language in its license offering a cash prize to the first user to notice the clause, it took nearly four months before someone collected the $1,000.[8]

When high-quality products are indistinguishable from poor ones, we get what economists call a market for lemons.[9] Even though car buyers would pay more for a vehicle with no mechanical issues, they often can't tell a reliable used car from a clunker destined to break down in a steaming heap in a week or two. And since they can't sort the good deals from the bad ones, they are only willing to pay a price corresponding to a low-quality car. But if buyers aren't willing to fork over extra money for a high-quality car, used car dealers have every reason to stock their lots with the cheapest cars available. So despite the fact that buyers would pay a premium for high-quality cars, the market fails to supply them.[10]

For the same reasons, most EULAs are lemons. Licensors have lots of information about what their licenses say. They drafted them after all. But the average person has very little information. This information asymmetry breeds disengagement and distrust. And if companies don't gain any advantage in the marketplace from more consumer-friendly licenses, that only serves to further entrench unfavorable terms. Once license terms are adopted, they have a way of spreading. In part, their viral nature is about saving time. Few licenses are drafted from scratch. Lawyers copy and paste liberally. When Google initially released its Chrome browser, the license read in part: "You give Google a perpetual, irrevocable, worldwide, royalty-free, and non-exclusive license to reproduce, adapt, modify, translate, publish, publicly perform, publicly display and distribute any Content which you submit, post or display on or through, [Google's products, software, services and web sites]."[11]

That would mean Google could publish every email you send, every photo you share, and every password you enter using Chrome. Of course, that's not what Google meant. It quickly updated the license to clarify that users retain copyright in the content they generate. Google's explanation for this mishap? "We try to use the same set of legal terms ... for many of our products. Sometimes ... this means that the legal terms for a specific product may include terms that don't apply well to the use of that product."[12] In other words, Google was a bit sloppy in copying from its existing licenses.

Somewhat less innocently, the uniformity in license terms is partly about safety in numbers.[13] Once a term becomes standardized, its inclusion becomes a strategy for reducing competitive risk. A company that adopts industry standard terms guarantees that it is no worse off than its competitors.[14] Combined with the lemon problem, this sort of soft collusion helps ensure that we don't see robust competition on the basis of consumer-friendly license terms.

Instead, we see a growing list of standard terms, almost none of which add to a product's value from the perspective of users. Some restrict what you can do with the products you purchase. These include limits on making backup copies, prohibitions on bad reviews, provisions permanently tying a product to a particular device, and bans on reverse engineering—the process of discovering how a product works through observing it in action.[15] Other terms eliminate legal rights and remedies. These include limitations on liability, bars to class action suits, and mandatory arbitration clauses.[16] And if the licensor neglected to include some one-sided term it later deems useful, many licenses give the drafter the option to change the terms of the EULA at any time.[17]

But for our purposes, the most important license provisions are ones that try to redefine ownership and limit the transfer of the products we purchase. Across the board, nearly every license agreement for digital content—software, games, music, movies, and books—declares that the product is licensed, not sold. As Apple informs its customers, "the software products made available through the App Store ... are licensed, not sold, to you."[18] Microsoft says the same thing: "We do not sell our software or your copy of it—we only license it."[19] Amazon's Kindle store follows suit as well: "Kindle Content is licensed, not sold, to you by the Content Provider."[20] Sony's PlayStation license states: "All Software is licensed, not sold, which means you acquire rights to use the Software ... but you do not acquire ownership of the Software."[21] The same sort of terms are increasingly attached to hardware devices with embedded software. Most of these licenses preclude you from reselling, lending, renting, or otherwise transferring their purchases.

There are two ways to interpret these kinds of terms. They might stand for the uncontroversial—and frankly obvious—proposition that when you buy a product like Microsoft Office software or an iPhone, you are not acquiring all of the copyright, patent, and trademark rights in that product. To which we say, duh. But these licenses typically mean something beyond that. They mean that you don't own the thing you buy. You don't own and can't transfer the plastic disc, the digital file, or the physical copy of the code embedded in your phone. So when retailers and record labels tell you that the song you purchased is licensed, not sold, they mean two things—you don't own the copyright in the song and you don't own the file you downloaded.

Despite this effort to define downloads of digital media as licenses rather than sales, rights holders take a very different position when it comes time to pay artists. Most recording contracts distinguish between sales—historically, of CDs and other physical media—and licenses—to use a song in a commercial, for example. A recording artist would be owed a small royalty for each sale, say 15 percent, and a much higher rate for each license, more like 50 percent. To minimize the payments owed to artists, record labels have insisted in lawsuits filed by Eminem and others that iTunes and Amazon transactions are in fact sales, not licenses.[22] Of course, that's not what their EULAs tell us. To understand how they benefit from that strategic characterization, we need to take a step back and consider how the licensing model first developed.

The Origins of the EULA

The EULA got its start in the software industry. In the early days of computing, hardware and software were typically bundled together. Software was a means of boosting hardware sales; markets for standalone software products had yet to develop. IBM was among the first companies to unbundle its hardware and software. But unlike its mainframes, its software code was easy to copy. IBM viewed existing intellectual property protections—patents, copyrights, and trade secrets—as either ineffective or too uncertain.

So it "coupled the copyright with a license and counted on the license to provide the real protection" against copying.[23]

At the time, IBM had legitimate concerns. Intellectual property law was decades away from unambiguous copyright or patent protection for software.[24] Even after the Copyright Act was amended in 1980 to explicitly recognize software as protected subject matter, developers worried about the dangers of rental. Since every PC is a copying machine, if Microsoft Word could be rented as easily and cheaply as a VHS copy of *Airplane!*, illicit copying could do real harm. At the urging of the software industry, Congress addressed that worry with the Computer Software Rental Amendment Act, which prohibited the rental of most software programs.[25]

But licensing is about more than guarding against legal uncertainty and the threat of infringement. Strategically, a business model premised on "licensing" products gave the software industry far greater control over downstream uses than other IP-intensive industries. Book publishers, record labels, and movie studios tried for decades to find a reliable and efficient means of stamping out unwanted competition by controlling secondary markets. By insisting that end users did not own the copies they bought, the software industry achieved that elusive goal and much more.

If you don't own a copy, you aren't entitled to resell or otherwise transfer it. That is just as true for software as it is for hardcover books. But because of the nature of software, ownership is even more crucial. Using software creates copies. If you install code on your hard drive, you've made a copy. If you run the program, you've created a copy in your computer's memory—at least according to some courts.[26] Unlike an analog book, a copy of a software program is virtually worthless without the right to make copies.[27] Congress understood that fact when it extended copyright protection to software in 1980. So it enacted a new broader set of exhaustion protections for software buyers.

Section 117 of the Copyright Act gives owners of copies of computer programs a number of important rights. First, it allows them to create "essential step" copies—copies necessary to run the program.[28] Second, it lets owners modify a program—to fix bugs, or add new features for example.[29] Third, it permits the creation of archival copies to guard against degradation and accidental deletion.[30] Section 117 also gives an owner expanded rights to transfer both the original copy they purchase and any unmodified essential step and archival copies.[31] The statute imposes two intuitive caveats on transfers. All rights to those copies must be transferred together as a single bundle. That is, the owner can't sell their original copy on eBay and their archival copies to a neighbor. And once the owner transfers those rights, any copies still in their possession have to be destroyed.

So long as the user is recognized as the owner, section 117 does its job. But by denying the existence of a sale, license agreements undermine these congressionally crafted consumer protections. If all you get when you fork over your cash is a license, you are at the mercy of the software maker. Your rights to transfer the copy, to make backups, even to install and use the software are determined by the text of the license, not by federal law or common sense.

Licenses displace the laws Congress created to secure consumer property rights. And by doing so, licenses make themselves an indispensable part of every software transaction.[32] If we accept the idea that the license prevents the transfer of ownership to the user, the license becomes the sole source of the rights to install and use the program. Without the license, the user can't do anything at all with their copy, aside from glare at it in frustration. Of course, if it weren't for the license's insistence that no sale has occurred, users wouldn't need permission to make those customary uses of the program in the first place. Those rights are already provided for in the Copyright Act.

In the software industry, these licenses have become commonplace, but these efforts to privately redefine consumer rights have spread to digital books, movies, and music—not to mention consumer electronics, home appliances, and farm equipment. As with software, the use, storage, or transfer of digital media products often requires the creation of additional copies. And just like software, these digital goods are encumbered by licenses that attempt to strip away property rights. But digital media aren't covered by the expanded statutory exhaustion rights that apply to software. So even if consumers can prove they own the digital files they acquire, they still need a license to make the copies necessary to use them. If ownership doesn't entitle you to actually use your property, ownership doesn't mean much.

If we accept the licensing model, we bow to private regulations that redefine consumer rights and impose conditions and restrictions on our use of the things we buy. Those constraints on our freedom aren't the product of self-governance. They are dictated by private actors driven by their own self-interest. The license agreement, in short, has us over the proverbial barrel.

But we shouldn't be so quick to accept the idea that license agreements define our rights. There are two ways of explaining the legal force of EULAs, both of which are flawed. First, we can treat them as contracts. But shoehorning contemporary consumer licenses into contract law requires us to toss out the basic rules and justifications for creating and enforcing contracts. More important, the contract model fundamentally misunderstands what's going on when rights holders claim they are licensing their products. The second approach recognizes that licenses aren't really about promises; instead, they are creatures of property law. But when we view licenses through the lens of property, it turns out rights holders often don't have the power that they assert over us. The notion that a license can strip buyers of their property rights is false.

The EULA as Contract

Most of us can remember signing a contract—closing on a home, signing a business loan, leasing a car, or perhaps entering into an employment contract. But very few of us would conjure up an image of installing software, downloading an ebook, or buying a new kitchen appliance as an example of a binding legal agreement. We tend to associate contracts with something more notable. Our common conception of contract formation involves substantial stakes and some formal process that clues us into the seriousness of the situation. There is probably a long document with lots of places to sign and initial. There might be someone walking you through the major provisions. Perhaps there are even lawyers in the room. But EULAs are of no moment. They are ubiquitous, an unnoticed part of the most mundane of modern tasks.[33] So we treat them like casual annoyances rather than binding obligations. Nonetheless, when courts are confronted with license agreements, they typically think about them as contracts.

Contract law reflects the deep moral intuition that we should be held accountable for the promises we make. By putting us on the hook for the harm we cause when we fail to live up to our word, the law encourages promise making and promise keeping. And when we can trust others to make reliable commitments, society works better. We can coordinate our actions, plan for the future, avoid costly protections against unscrupulous behavior, and preserve valuable relationships.

But it only makes sense to hold people to their word when they know that they are entering into a binding agreement and understand the terms. Traditionally, contract law had built-in mechanisms to make sure contracts reflected the mutual intent of the parties. But those checks on contract formation are broken. Today, many courts are willing to enforce terms that consumers do not understand, did not read, have never seen, and to which they simply didn't agree, so long as there is some constructive notice of their existence. It is this mutant form of contract law that embraces the EULA.

Under the classic model, forming a contract requires mutual assent to the terms by the parties—typically established through an offer by one party and its acceptance by the other—and what the law calls "consideration." To form the basis of a contract, an offer has to be definite and needs to reflect all of the terms central to the deal. Imagine your neighbor rings your doorbell and proposes to "sell you some stuff." Appealing though it may be, that suggestion doesn't function as an offer. You can't sensibly agree to such an amorphous deal. At the very least, you'd need to know what stuff and at what price.

Then your neighbor clarifies, "I'll sell you my creepy antique mannequin collection for $500." Fan though you are of moldy, dead-eyed, department store dummies, the price strikes you as steep. You reply, "You've got a deal for $350." Even though you expressed your willingness to buy the collection, this is not an acceptance of your neighbor's offer. Acceptance requires agreement on all material terms like price and quantity.

Assuming you agree on all the important points of the deal, acceptance of a contract can take many forms. You and your neighbor could type up the terms and sign them. But most contracts can be formed orally, no writing required. Contracts can also be accepted through your actions. Your neighbor could say something like, "Think it over. If you agree, leave the money in my mailbox." By dropping off the payment, you'd be accepting the terms of the deal.

Offer and acceptance are important because they provide strong evidence of mutual assent. A contract should reflect a "meeting of the minds" between the two parties. They should have a common understanding of what they are each required to do.

Finally, a contract requires consideration—something of value provided to induce the other party's participation. So in our antique mannequin example, your neighbor's consideration is the promise to transfer their collection to you. And your consideration is payment of the agreed-upon price. Imagine instead that your neighbor, in a generous mood, stops by and says, "I promise to give you my mannequin collection tomorrow." And you reply, "Sounds great." Since you've provided no consideration—you haven't obligated yourself to do anything—there's no contract. Your neighbor is free to change their mind tomorrow. It's your consideration that obliges the other party to hold up their end of the bargain.

When it comes to EULAs, many courts have essentially abandoned the traditional rules of contract formation. They enforce terms we find inside a package only after making a purchase. These terms don't require us to explicitly agree to anything. Simply using the product or even opening the packaging is enough to bind us. Then there's the now ubiquitous "I Agree" button. Of course, clicking that button is no guarantee of meaningful assent since almost no one reads the terms before reflexively "agreeing" to them. But courts enforce them anyway. Some courts have even gone so far as to bind parties to terms linked from a website, regardless of whether they ever saw them. Assent, such as it is, is manifested by merely visiting the site.

When parties don't see the terms of an agreement until after they make a purchase, when they never see the terms at all, or when they take no intentional steps to manifest assent, contract formation rules are stretched to their breaking point. As a result, a number of early courts refused to enforce EULAs.[34] But over time, courts grew to accept them. As Mark Lemley has explained, "A majority of courts now reject any requirement that a party take any action at all demonstrating agreement to or even awareness of terms in order to be bound by those terms."[35]

Judge Frank Easterbrook's opinion in *ProCD, Inc. v. Zeidenberg* is largely to blame for this distortion of contract law. ProCD sold a CD-ROM database of telephone listings to most customers for $150; it sold copies to retailers and other commercial users at a significantly higher price. In order to maintain this price discrimination strategy, ProCD included a license with the low-cost version that, among other things, prohibited commercial use of the database. The product packaging noted that it contained a license, but users had no opportunity to review its terms until after they purchased the software. Matthew Zeidenberg bought a copy of the database and posted it online, charging users a fee to access it. ProCD sued. Since copyright law does not protect purely factual compilations like lists of names and phone numbers, ProCD's claim was based on a breach of contract. The question for the court was whether the noncommercial use restriction was part of the agreement between ProCD and Zeidenberg.

For most consumer goods, the contract is formed at the time of initial purchase. Let's say you walk into your local hardware store to buy a shovel. You see one that looks suitable for your needs. It bears a $20 price tag. Even though it's not as formal as a loan document, contract law calls that an offer. You take the shovel to the checkout counter and tender the asking price. That's acceptance. A contract has been formed. But let's say that once you get home, the hardware store calls you and says, "You know that shovel you bought? Well, there are some additional strings attached. You can use it for ditch digging, but you can't use it for gardening. Gardening requires you to pay an extra $30 upgrade fee." After depleting your reserve of expletives and hanging up the phone, chances are you would feel no obligation to avoid planting some shrubbery with your new shovel. And no court in its right mind would disagree with you. It would recognize that phone call as an ineffective attempt to modify an existing contract.

Now imagine a slightly different scenario. While waiting in the checkout line, you notice a sticker on the handle of your shovel. It reads: "This shovel is subject to a license agreement. You will be notified of the full terms by phone after your purchase." Does this change things? Most of us would probably say no. Vague references to unknown terms cannot form the basis of a contract. But according to Judge Easterbrook, the fact that Zeidenberg was put on notice that the license terms were forthcoming, even if he had no idea what they were at the time of purchase, was enough to make them part of a binding legal agreement.

Easterbrook offered a number of justifications to soften the blow of his departure from the basic rules of contract formation. But rather than putting our minds at ease about enforcing these kinds of terms, each of Easterbrook's assurances underscores the legitimate worries of consumers. First, he suggests that if the license agreement is a contractual offer, you can always reject it. If you don't like the terms, you can simply return the software to the store. Anyone who has ever tried to get a refund from a retailer for opened software products can point out the obvious flaw in this reasoning.

Beyond these optimistic predictions about return policies, failure to take action is a problematic trigger for contract formation. If one day a neighbor walked up to you and said, "I propose to buy your house for $1. Failure to mow my lawn by tomorrow constitutes acceptance of my offer," we would be shocked if a court called that an enforceable agreement. But that's just what the court does in *ProCD*. Your failure to return the software constitutes acceptance. In treating inaction as assent, Easterbrook ignores the costs this sort of arrangement imposes on users. Under his view, software transactions require you to drive to a retail store, find the software you want, pay for it, take it home, and inspect the terms. If you don't like the terms, you have to drive back to the store, wait in line for a refund, explain why the box is open, and hope that the manager at the local big-box office supply store is willing to make an exception to its refund policy.

This reality is at odds with one of Easterbrook's chief defenses of enforcing EULAs—market efficiency. By standardizing agreements, the argument goes, we streamline the process of mass production and distribution. To go back to a requirement of individualized contracts, he says, would "return transactions to the horse-and-buggy era." Standardized mass contracts, in contrast, hold out the promise of drastically reducing transaction costs for sellers.

Easterbrook is right that standardized contracts lower costs for software makers. They draft one license, likely cobbled together from existing terms, and use it in thousands or even millions of transactions. No messy negotiations, no discussions, no explanations. Undoubtedly, that reduces costs within the software industry. And while it is generally true that reducing transaction costs is a good thing, here those costs are not eliminated. They are just shifted from sellers to buyers. In a world governed by EULAs, life is easier for software companies and much harder for all of us. We are the ones expected to read and understand page after page of license text. And those costs add up. The failure to account for them shows that Easterbrook is keenly concerned with transaction costs when they harm software makers, but remarkably insensitive to those costs when they are imposed on individuals.

Next, Easterbrook gestures toward competition as a check on abusive license terms. If people are unhappy with a term that restricts how they can use a product, he speculates, surely competitors will offer more attractive terms to win them over. But the information asymmetry between users and license drafters makes it unlikely that the market will reflect consumer preferences. For the average user, the costs of researching license terms far outweigh the value of the goods at issue. The same is not true for the seller, who has the power to make certain that its license reflects its own best interests and has very strong incentives to do so. That dynamic all but ensures that competition will not result in more favorable terms. In fact, there is good reason to expect competition will lead to worse terms as companies look for ways to keep prices—the most obvious point of comparison—low.[36]

But even if competitive forces don't weed out unfair terms, Easterbrook reassures us, contract law will. The doctrine of unconscionability will prevent enforcement of EULAs if they are truly egregious. A contract is deemed unconscionable, and thus unenforceable, when one party enjoys superior bargaining power and the substance of the agreement is so one-sided that no reasonable person would agree to it. That's a tough standard to satisfy. This rule looks at both the process by which the contract was formed and its substantive terms to decide whether the apparent agreement of the parties should be set aside. In terms of process, take-it-or-leave-it contracts like EULAs, sometimes called contracts of adhesion, typically suggest unequal bargaining power. But by lending its enthusiastic seal of approval to the pay-now, terms-later EULA, the *ProCD* opinion makes the uphill battle to establish unconscionability even more difficult.

Finally, Easterbrook reminds us that contracts don't create rights against the world; they only create rights as between the parties to the agreement. So even if these contracts are enforced, their impact will be limited. In contrast to public law like property or copyright, a contract does not affect the rights of the public at large. As Easterbrook explains, "Someone who found a copy of SelectPhone on the street would not be affected by the shrinkwrap license."[37] In a formal sense, it is true that only parties to the contract are bound by it. But once we accept the distorted picture of contract formation endorsed by *ProCD*, we all become parties to these contracts, as even Easterbrook's own example shows. The stranger on the street who finds a copy of ProCD's product may not be bound the moment they pick it up. But once they install the program, "the software [will splash] the license on the screen and [will] not let him proceed without indicating acceptance."[38] Everyone who encounters the product is restricted in their use of it. Those restrictions travel with the product, much like the hypothetical restrictions on tuxedo rental forbidden by property law.

In the two decades since *ProCD*, EULAs have flourished. For the vast majority of us, these ubiquitous license terms are unnoticed, unread, and unreadable. The practice of entering into these so-called agreements has become automatic. We unthinkingly click "I Agree" when we buy a product online, when we download a new app on our phones, and when we log onto our online banking accounts. And thanks to lax standards for contract formation, we enter into binding legal obligations simply by visiting a website that links to a set of terms. By announcing their intention to bind you, the operator of any website can rope you into a contract. What used to require a meeting of the minds is now a unilateral exercise of power. Even after these one-sided agreements are formed, many allow the drafter to change the terms at any time without assent.

Under these circumstances, it is hardly surprising that people don't bother reading EULAs. Typically, the failure to read a contract is no excuse for breaking it. The law imposes a duty to read, and for good reason. When people know they are entering into an agreement, generally the reasonable thing to do is read it. Closing your eyes and plugging your ears doesn't get you off the hook. But the duty to read doesn't look quite so reasonable when the consumer has little reason to expect a contract is in the offing. And even when they do have reason to know a contract has been presented—when they confront the "I Agree" button—the duty to read should take into account the cost of studying the terms presented. If the duty to read is about making sure people behave reasonably, we might ask whether it is in fact reasonable to expect a consumer to read a *Macbeth*-length license before they make a 99-cent purchase.

So despite its widespread acceptance by courts, the notion that EULAs are enforceable contracts rests on a shaky foundation and leads to a slew of troubling consequences. But there's another way to think about the license agreement, one that more accurately captures how they function.

The EULA as Permission

If EULAs aren't contracts, what are they? The classic understanding of a license has nothing to do with mutual promises. A license is a pure expression of permission, one that requires no agreement to be effective. Let's say you are walking along a secluded country road and notice a picturesque lake in the distance. You decide to rest in the shade of a lakeside oak tree. But as you make your way closer, you see a sign that reads "No Trespassing." Despite your vigorous disagreement with being denied a peaceful afternoon in the shade, your assent, or lack thereof, is utterly irrelevant. You and the property owner don't have to agree that you keep out. The owner's assertion is all that matters. The same would be true if the sign read "Feel free to enjoy the shade, but absolutely no swimming." The moment you take a dip, you become a trespasser.

EULAs work in a similar way. The owner of an intellectual property right can define the circumstances under which others are allowed to use their creation. Sometimes they grant permission, other times they don't. And even though that permission might be memorialized in a written document, mutual assent isn't required for the limits on that grant of permission to be enforceable.[39] Let's say you write a hit song. General Motors comes to you and asks to put that song in a commercial. You tell them, "You can use my song, but only for ten seconds. And definitely not in a Buick ad." If they use a twelve-second clip of the song in a Buick Enclave commercial, they've infringed your copyright regardless of whether a contract was formed or not.

Nonetheless, most courts, commentators, and copyright holders continue to think of licenses as creatures of contract law. The free software movement is one notable exception. Developers of free software are committed to the idea that all users should be free to run software, study it, modify it, and redistribute it. Those core beliefs are reflected in free software licenses like the GNU General Public License, or GPL. Examples of free software products include the Firefox web browser, the Apache web server, and MySQL relational database software. As Eben Moglen, head of the Software Freedom Law Center and one of the drafters of the current version of the GPL explains, "Licenses are not contracts: the work's user is obliged to remain within the bounds of the license not because she voluntarily promised, but because she doesn't have any right to act at all except as the license permits."[40]

An approach that roots licenses in property law is preferable to one that treats them like contracts. It does away with the fiction of assent. That could make it easier for licensors to assert their wills, but it would also avoid the damage done to contract law by insisting that EULAs are enforceable agreements. So why don't we see more rights holders adopting this stance? One answer is path dependence. *ProCD* illuminated a clear and not particularly arduous path for enforcing license terms. Most rights holders—and their lawyers—are too risk-averse to rely on a sounder but largely untested argument.

The license-as-permission approach also helps clarify the relationship between the license and property ownership. A license does not—in fact, cannot—define property rights; it depends on predetermined property rights. A license is a tool that allows a property owner to control how others use a resource. Before a license can be effective, we have to know who owns what. If the party asserting the license doesn't own the property, the license is an empty gesture. If you post a "No Trespassing" sign in a city park, for example, it has no legal effect since that's not your property in the first place. If you don't own it, you can't license it. And the license itself cannot transfer ownership from one party to another. So before a valid license can exist, we have to know who owns the resource.

The requirement that the licensor have some property or statutory exclusive right is one reason the court in *ProCD* insisted that EULAs are contracts. ProCD's database of phone listings did not qualify for copyright protection; it was part of the public domain. As a result, ProCD had no legal right to exclude others from the information. It had no property to license. Contract law was the only way to impose restrictions on the use of what was essentially public property.

You can't use a license to prohibit joggers in the city park. Likewise, copyright holders can't use licenses to control behaviors that aren't within the scope of their statutory rights. But that doesn't stop them from trying. They try to prohibit things like negative reviews or reverse engineering—uses that are noninfringing and thus outside of their control. Those kinds of licenses are more like a "No Trespassing" sign that forbids you from describing the roadside lake to a friend. The lake's owner has property rights, but they just don't extend that far.

There are plenty of examples of valid copyright licenses. When a copyright holder grants permission to make a derivative work—say a novelist allows a movie studio to adapt a book into a film—that is squarely within the scope of the novelist's statutory rights. So are grants allowing the public performance of a play or the reproduction of a photograph in an ad campaign. Copyright holders can also license the distribution of copies that they own. If an artist lends—rather than sells—a sculpture to a museum, they can prohibit the museum from lending it to another institution. When they grant permission over uses within their statutory rights, copyright holders can give, withhold, or condition permission in all sorts of ways.

But not all rights related to a work belong to the copyright holder. Some are reserved for the public at large. Others are granted to owners of copies. One crucial function of the Copyright Act is to divvy up rights between creators and consumers. Without copyright law, the public could copy, distribute, and adapt every book, record, and film produced. But copyright law takes those rights away from the public and gives them to creators and their publishers. The public still has the right to make fair uses of protected works.[41] And eventually copyright expires, works enter the public domain, and the public is free to use them. But copyright law also carves out a set of rights for owners of copies of protected works. That's what the exhaustion rules embodied in sections 109 and 117 do. The law says that people who own copies have rights that the public at large doesn't. They can distribute their copies of books by reselling or lending them. They can publicly display their paintings. They can make backups and adaptations of their computer programs. Those are rights copyright holders would strongly prefer to control, but the Copyright Act gives them to purchasers instead.

By insisting that their products are licensed rather than sold, copyright holders are trying to get those rights back. The courts and Congress have decided certain rights belong to us, but the license is designed to overcome this default allocation of property rights. On the one hand, if we think of licenses as contracts, this transfer of property rights from copy owners to copyright holders makes some conceptual sense. Parties agree to transfer property rights all the time. You might agree to sell your house for a hefty sum, but that transfer depends on the mutual agreement of the parties. If, on the other hand, we think of a license as a pure expression of permission, a license attempting to reclaim exhaustion rights is about as effective as your neighbor declaring, over your strenuous objections, that they now own your spare bedroom. It just doesn't work.

When it enacted the exhaustion provisions in the Copyright Act, Congress intended to safeguard personal property rights. If that was the goal, it would be a very odd choice to allow copyright holders' unilateral edicts in the form of EULAs to eliminate those rights. It strikes us as highly unlikely that Congress went through the trouble of allocating rights to owners if those rights would ultimately be contingent on the kindness of copyright holders to refrain from reclaiming them.

Copyright holders assume that personal property rights are theirs to grant or withhold. And on some level, that is true. A copyright holder can choose not to release a work to the public at all. They can leave it in a vault collecting dust. Or they can choose to exhibit it publicly, but refuse to sell individual copies, as hip-hop group the Wu-Tang Clan did with a recent album that required fans to visit a museum to hear the single copy produced.[42] Copyright holders can choose to lease or rent their works to the public, but not sell them. But whether or not a transaction is a sale that transfers ownership can't be up to the rights holder alone. Understood properly, licenses that attempt to redefine consumer property rights in their purchases fail.

In part, that's because property law puts limits on the kinds of transactions the law will recognize. You can enter into a contract that obligates you to refrain from renting your tuxedo, but you can't sell a tuxedo with a no-rental restriction that is enforceable against the world. Allowing those kinds of idiosyncratic restrictions defeats one of the main benefits of personal property law—clear legal rules that reduce information costs. The economy would grind to a halt if every consumer transaction required a diligent investigation into the strings attached to every apparent sale. Instead, we need some objective basis for determining whether consumers are owners or not. But courts have struggled mightily to come up with a workable test for identifying sales.

Defining Ownership

So how can we tell whether or not a reader, listener, or user owns something? Too many courts, especially in cases dealing with software, have turned to license agreements, on the assumption that what they say goes. As long as the copyright holder recites some variation of the magic words "this is a license, not a sale," you don't own anything.[43] The better approach, which a number of courts have adopted, turns to some source of publicly made law. They might look to centuries-old common law property rules, or the rules governing the sale of goods outlined in the Uniform Commercial

Code,[44] or the internal rules of intellectual property law. These sources focus on objective facts about a transaction, not just the self-serving claims made in the license agreement.[45]

The Copyright Act, for example, grants copyright holders the exclusive right "to distribute copies ... to the public by sale or other transfer of ownership, or by rental, lease, or lending." This language suggests that copyright law recognizes two kinds of transfers of copies. First, we have permanent transfers: sales or gifts. Second, we have temporary transfers: rental, lease, or lending. The first are transfers of ownership; the second are not. The question then becomes which of those two categories is a better fit in a particular circumstance. Plenty of transactions are easy enough to characterize as a rental, lease, or lending. You don't own the books you borrow from the library. And your Netflix subscription doesn't give you a property interest—personal, intellectual, or intangible—in the movies you watch. But when you pay a one-time fee for a copy of, or permanent access to, an ebook, game, or other digital media, that should be recognized as a sale that transfers ownership of a tangible or intangible asset.

Courts struggle to define and identify sales in large part because they can't decide whether to rely on the privately drafted declarations of copyright holders or facts about a transaction beyond the license. There is no better example of this floundering than a pair of cases argued on the same day in front of the same three-judge panel of the Court of Appeals for the Ninth Circuit, a court whose territory includes both Hollywood and Silicon Valley. Both cases involved the resale of copies despite license terms prohibiting transfer. And both cases turned on the question of copy ownership. If the defendants owned their copies, they were free to resell them. But if the copies were licensed, reselling them was an act of infringement. After years of inconsistent decisions, many hoped these cases would clarify the question of consumer ownership. Although the license agreements in both cases imposed nearly indistinguishable restrictions, the judges reached very different conclusions, relying on two incompatible approaches. In one case, ownership was determined on the basis of objective evidence about the nature of the transaction. In the other, the court relied solely on the pronouncements of the copyright holder.

UMG v. Augusto involved the resale of promotional CDs. Record labels frequently send free CDs to critics, bloggers, and other tastemakers. Inevitably, those CDs end up for sale at used record stores. Troy Augusto made his living buying used CDs—including the relatively rare and profitable promo CDs—from local record stores and reselling them on eBay. Augusto figured that since he owned the discs, he was entitled to resell them. But

Universal Music Group insisted that he didn't own the CDs because of a license printed on the discs. That license claimed that the discs remained the label's property, limited recipients to noncommercial use of the discs, and prohibited resale and transfer. Nonetheless, the Ninth Circuit held that ownership of the discs transferred to their recipients upon delivery and, eventually, to Augusto. Although it cast some doubt on whether the notice was enough to form a binding agreement, the court focused on Universal's method of distribution. Universal made no effort to keep track of the CDs it claimed to own once they were shipped. After dropping them in the mail, it had no control over how they were used or by whom. And Universal had no means of collecting the CDs it claimed it owned. The discs were under the control of the recipients, who were free as a practical matter to use them as they pleased.[46]

The other case, *Vernor v. Autodesk*, centered on the resale of software discs. Like UMG, Autodesk argued that the notice accompanying its software meant that end users who paid thousands of dollars for copies did not own those plastic discs. Autodesk merely "licensed" them. Rather than considering the kinds of factors they relied on in *Augusto* however, the judges created a three-part test that asked (1) whether the copyright holder called the transaction a license; (2) whether it restricted transfer of the software; and (3) whether it restricted use of the software. Since Autodesk's license terms contained the necessary language, the court concluded that Autodesk, not the end user, owned the discs.

The *Vernor* test is flawed.[47] It hinges entirely on self-serving proclamations from the copyright holder. By reciting the appropriate magic words, a rights holder can avoid a sale regardless of the objective reality of the transaction. Even if you pay a one-time price for an item you get to keep forever, as long as the license repeats a few key phrases, no sale has occurred. The test also begs the question. The reason we need to decide if there's been a sale is to know whether the buyer can transfer their property over the objection of the copyright holder. Under the court's test, copyright holders can defeat the buyer's property claim by objecting to resale and lending. But to an owner, those objections are irrelevant; so they don't help us answer the question of ownership. Again, that's because licenses depend on clear property rights; they don't define those rights.

We think there is a better way to answer the ownership question—one that is more accurate, reliable, and fair. As other courts have recognized, the economic reality of a transaction is the best guide to deciding whether a sale has occurred.[48] There are three considerations that offer strong indications of ownership: (1) the duration of consumer possession or access; (2)

the payment structure of the transaction; and (3) the characterization of the transaction communicated to the public.

Under the first factor, we follow the lead of the Court of Appeals for the Second Circuit, which said it would be "anomalous" to treat a user as anything less than an owner when their "degree of ownership of a copy is so complete that he may lawfully use it and keep it forever, or if so disposed, throw it in the trash." Under the second factor, someone who pays a one-time fee is more likely to be an owner than someone whose access depends on ongoing payments. If you need to pay a monthly fee to access a collection of movies, for example, you are a subscriber, not an owner. Under the third factor, we take into account the way a transaction is presented to the public. If a service is clearly advertised as a subscription, for example, ownership is harder to argue. The hard question here is what communications count. Copyright holders want their license terms to be dispositive. But as we all know, people rarely read them. More importantly, the fine print is often overshadowed by the simple language used to market digital media. Apple, Amazon, and others implore you to "Buy Now," "Purchase," and "Own It in HD." We think those kinds of statements are more important in establishing consumer expectations and should figure heavily in the determination of consumer rights.

If courts consistently thought through these considerations, they would reach conclusions about ownership that were fairer and more intuitive. More of us would own the products we buy and enjoy greater freedom to use and transfer them. It would also mean that rights holders would have a harder time configuring bespoke bundles of rights. That loss of flexibility could have an impact on price. Without the ability to tailor licenses, they argue, it will be more difficult for rights holders and retailers to tailor their prices. We interrogate those claims below.

Licensing and Price Discrimination

Licenses facilitate price discrimination. *ProCD v. Zeidenberg* illustrates this point well. ProCD wanted to sell its database to two different groups of customers at very different prices. Businesses like telemarketing companies were willing to pay high prices for ProCD's database. But the average person has less money to spend and less interest in a phone database. So the price had to be lower. If ProCD charged a high price, businesses would buy, but normal people wouldn't. If it charged a low price, both would buy, but ProCD would be leaving money on the table since businesses would have paid the higher price. The solution is price discrimination. Charge businesses high prices, and charge the average user low prices. By doing so, ProCD can maximize its profits.

From a seller's perspective, the ideal world would look something like this. Information about each potential customer's preferences, needs, buying habits, bank account, physical condition, and emotional state would give the seller real-time information about exactly how much they are willing to pay for a particular product. Running late? Expect higher gas prices. Parched after a long run? Expect to pay twice as much for that bottle of water. Just got paid? Phone battery about to die? Expect your Uber to cost twice as much. Expect those new shows you've been eyeing to cost a few dollars more. Despite the terabytes of consumer data collected in servers across the globe, the dream—or nightmare—of perfect price discrimination isn't yet a reality. But we seem to be on our way.[49] Google recently patented a technology that allows it to predict—leveraging the massive digital dossier it has compiled over the years—how likely a customer is to buy a particular product and adjust prices accordingly.[50] Not to be outdone, Facebook patented a technology that helps lenders discriminate based on borrowers' social connections.[51]

But there are other, less precise ways to discriminate. Sellers can offer slight variations of the same product at differing prices. Bulk sales are one familiar example. Buyers who are price sensitive can get a year's worth of ranch dressing at their local warehouse store. And those with more money to spend can buy a single bottle. Or think of airlines. Coach and first class passengers get essentially the same service—transportation from one city to another. But first-class passengers pay for bigger seats, better food, and more personal service. And airlines don't just divide customers into coach, business, or first; United, for example, relies on over twenty different fare classes, each defined by its own sets of perks and restrictions—and each with its own price.

Sellers can also discriminate between different groups of consumers, determining prices on the basis of various demographic proxies for willingness to pay. That's what ProCD did when it divided the world into commercial and noncommercial users. That's what John Wiley tried to do when it divided the college textbook market into the United States and everyone else; what restaurants do with early bird specials; and what movie theaters do when they offer student discounts.

Resale can disrupt these carefully laid plans. Price discrimination depends on the ability to limit arbitrage—the practice of buying goods at low prices in one market, and selling them for more in another. If individuals can buy cheap copies of ProCD's database and sell them to commercial users, ProCD misses out on potential revenue. That's one reason rights holders prefer licenses to sales. With a license, there is no exhaustion, no property rights, no resale, and no arbitrage. They can be sure that their preferred pricing scheme won't be undermined by enterprising resellers. The question isn't why sellers want to put a stop to resale markets, it's whether we should let them.

Advocates of price discrimination argue that it benefits consumers, or at least that it can. First, price discrimination—the argument goes—can keep prices low by requiring wealthy buyers to subsidize more price-sensitive shoppers. Second, it creates incentives for product differentiation that increases consumer choice. We think there are good reasons to be skeptical of both of these theoretical upsides. We acknowledge that there are times when particular subsets of consumers benefit from price discrimination. But on the whole, it is a strategy that transfers money and control from the public to rights holders.

Let's think about price first. How does price discrimination keep prices low? Without price discrimination—as Judge Easterbrook argued—ProCD could be forced to raise its prices. Instead of selling its low-cost product for $150, it might have to raise the price to $200 to account for lost revenue from higher-priced sales to commercial users. This is essentially the same argument John Wiley made to the Supreme Court; if it couldn't discriminate against U.S. students through high prices, it would be forced to raise the prices for Thai students. These examples both show how price discrimination can help relatively poor consumers at the expense of wealthier ones. Assuming you support this redistribution of wealth, you might question whether these implicit subsidies should be entrusted to private actors or if instead they should be crafted through public debate and decision making.

Nor is it the case that price discrimination always favors the poor at the expense of the wealthy. The market for consumer credit, for example, is just the opposite. For the wealthy, credit is cheap and convenient. For the poor, it's anything but. Credit is expensive, and subsistence credit card users are among the industry's most profitable customers.[52] Mortgages, home loans, and even groceries cost more for poor families than rich ones.[53] Ultimately, which group benefits from price discrimination—rich or poor, domestic or foreign, young or old—depends on what's best for the seller. *ProPublica* recently reported that Asian families are charged higher prices by the test prep company Princeton Review, for example.[54] And a recent White House report noted that, as online intermediaries gather more information about us, these practices "raise the specter of 'redlining' in the digital economy—the potential to discriminate against the most vulnerable classes of our society under the guise of neutral algorithms."[55]

First and foremost, price discrimination is a strategy for maximizing the seller's profit at the expense of buyers. Imagine a market without price discrimination, one in which every buyer pays the same price for a given product. In that market, let's say this book costs $20. Readers who value it at less than $20 won't buy it. Don't worry, we won't be offended. Those that value it at $20 or more will buy it. Let's say you think this book is worth $25. When you buy it for $20, you realize a $5 surplus—the difference between your personal valuation of the book and the price you actually paid. Across the economy as a whole, that surplus is worth trillions of dollars to consumers.

To sellers, that surplus represents untapped revenue. The goal of price discrimination is to reduce the consumer surplus to zero. If you value the book at $25, that's what you pay, and not a penny less. As we are divided up into smaller groups, prices can be more carefully tailored to match our willingness to pay. When that happens, the consumer surplus goes down and sellers come away with more of our money. As the economist Louis Phlips put it, "Price discrimination aims at taking the entire consumer surplus away from all customers, if possible."[56] So to the extent we benefit from price discrimination, sellers see that fact as a bug, not a feature. Not surprisingly, most people are wary of price discrimination. A University of Pennsylvania study found consumers overwhelmingly object to price discrimination, viewing it as morally wrong and legally suspect.[57]

That's not to say we shouldn't care about affordability and making products available to those with fewer resources. But we think the secondary markets for resale and lending that exhaustion makes possible are a better way to achieve that goal. Secondary markets keep prices down more efficiently, more reliably, and without the collateral damage of price discrimination.[58]

The second argument in favor of price discrimination is that it increases the number of options in the marketplace. Because it encourages product differentiation, price discrimination results in increased consumer choice. Think of the array of options available on a new car. Each accessory you add creates a slightly different product, one customized for you. And it's not just big ticket items. Nike and Converse now let shoppers customize every square inch of their shoes.[59] You can even buy M&Ms bearing the message or image of your choice. Custom products aside, a trip down any aisle at the grocery store reveals the staggering number of choices. Crest, for example, sells forty-one varieties of toothpaste in the United States alone.[60]

When we make important decisions—which house to buy or what college to attend—an ample supply of choices can be helpful. But for more mundane choices, an overabundance of options leaves us in a market that is more difficult to navigate.[61] The surfeit of product features, ingredients, sizes, and quantities—each with their own impact on price—can undermine our ability to draw comparisons between products and make informed decisions. By increasing the cost of information, customization can make it harder to know whether you are paying a fair price and increases the risk of buyer's remorse. Licenses are a perfect vehicle for this sort of price discrimination. They allow nearly infinite flexibility to craft whatever combination of rights a seller can imagine. If there's reason to think some consumers will pay more for a license that lets them read their ebooks in the bathtub, it's easy enough to capture that surplus. And since license terms are rarely read or fully understood, they reinforce the opaque nature of pricing.

We are not opposed to consumer choice. In fact, we think meaningful options are crucial to the functioning of markets for digital goods. But there comes a point at which additional choices do more harm than good. For obvious and important product attributes—like whether you are buying a movie directed by Jean-Luc Godard or one starring Jean-Claude Van Damme—the more choices, the better. But when it comes to nearly indistinguishable variations between license agreements, additional options harm the public on the whole. We aren't suggesting a one-size-fits-all solution that requires everyone to own the media they consume and the devices they use. We do think, however, that rental and subscription models offer clear, easy to understand alternatives and considerable flexibility. Netflix, for example, charges one price for individual users and another for its family plan. And unlike licenses, rentals and subscriptions don't result in ambiguity about whether a sale has occurred. As we detail in the next chapter, licenses create considerable uncertainty about precisely what we get for our money.

5 The "Buy Now" Lie

If you've read this far, you understand the potential disparity between the legal rights of purchasers of analog and digital goods. The owner of a hardcover book, for example, can lend, resell, or give away their purchase consistent with age-old principles of personal property. But according to digital retailers and publishers, an ebook owner can't do the same. Nonetheless, the market for digital goods continues to grow. What explains the apparent eagerness of consumers to sacrifice these economically and socially valuable rights?

Ordinarily, the free market sends strong signals about what consumers want. Those signals tell companies which products to make, how many to produce, and what they should cost. So we would expect the market demand for digital goods to accurately reflect changing preferences. If we are buying more digital goods, that suggests the convenience and other advantages of those products outweigh whatever value people assign to ownership.

But the signals the market sends are not always reliable. One foundational assumption of market economics is that consumers make choices based on accurate information. But if consumers are denied valuable information—about competing products or current prices, for example—their behavior in the marketplace is a less useful indicator of their preferences. Information is never perfect, but we have laws that are intended to guard against the most egregious sources of misinformation. Trademark law, for example, prevents the use of confusingly similar names, logos, and other indicators of the source of a product.[1] More directly, the law prohibits false and deceptive advertising.[2] A company can't claim that its bottled water cures cancer if it doesn't, or that its service is free if it isn't.

Some people—and we are occasionally among them—prefer digital goods despite a full understanding of the limited bundle of rights associated with them. But not everyone is so well informed. Only a tiny fraction actually read the fine print that spells out the details of digital transactions. And even if they did, most would struggle to make sense of those terms. Equally troublingly, those terms appear to conflict with a common-sense understanding of other, more prominent messages used to market digital goods. Words like "buy" and "own" are casually and commonly employed by digital retailers. And to the unwary, those words could communicate a set of rights that corresponds to personal property in the analog world. If so, digital retailers are engaged in a pervasive false advertising campaign that could have wide-reaching consequences for our shared understanding of ownership.

Mixed Signals

The digital marketplace is rife with marketing language that makes promises about ownership that are inconsistent with the text of the licenses that retailers insist govern these transactions. A shopper browsing digital movies on the Apple iTunes Store, for example, is likely to run across an ad inviting them to "Own It in HD." Unsurprisingly, Apple's marketing materials do not define precisely what it means to own a movie purchased from its digital storefront; it leaves customers to fill in those blanks for themselves. People with a lifetime of experience owning tangible objects could be forgiven if they assumed that the same basic rules of personal property applied to iTunes purchases. But Apple's license, despite describing those transactions as "purchases" and noting that "all sales ... are final," insists that customers cannot "rent, lease, loan, sell, [or] distribute" the movies and music they acquire from iTunes.[3]

Or consider Amazon, which offers hundreds of millions of items for sale on its site, from books and CDs to treadmills and toupees. Amazon invites its customers to purchase each of these millions of items with the ubiquitous phrase "Buy Now," or if you've enabled single-click shopping, the somewhat awkward "Buy Now with 1-Click®." That's true whether the product is a physical object or digital file. A reader shopping for an ebook encounters the same invitation to "Buy Now" they would see if contemplating a hardcover. And they would complete that transaction by clicking the very same button.

Despite these similarities, Amazon offers fundamentally different products to buyers of digital and analog goods. Those who buy hardcover books own their purchases; those who "buy" ebooks have a different relationship with their books, one that is probably unfair to characterize as ownership. Amazon says that relationship is defined by its terms of service. And buried within the thousands of words of that EULA is one consistent message: you don't own your ebooks; you merely license them. You have permission to use them in the ways Amazon permits, and that's all. As Amazon's terms explain, "[u]nless specifically indicated otherwise, you may not sell, rent, lease, [or] distribute ... any rights to the Kindle Content."[4] Amazon's MP3 store offers similar terms, albeit in a separate prolix document. Although Amazon customers "purchase" music, payment merely "grant[s] you a non-exclusive, non-transferable right to use. ... Music Content ... only for your personal, non-commercial purposes." And "you may not redistribute, ... sell, ... rent, share, lend, ... or otherwise transfer or use Purchased Music."[5]

Sometimes ownership is presented as an explicit selling point of digital content despite obvious limitations on the rights of buyers. When publisher Image Comics announced a digital storefront for comic books, it touted the fact that unlike competing digital comics services, customers actually owned their purchases. *Wired* published an article with the headline "For the First Time, You Can Actually Own the Digital Comics You Buy," reporting on the difference between the Image Comics site, which allows customers to download DRM-free comics to their hard drives, and competing services, which prohibited downloads.[6] As Image's Director of Business Development explained at the time, "There's something to be said for the ownership factor. If readers purchase a book on ComiXology, ... that could be revoked. And God forbid, if ComiXology goes under or their data center has an earthquake all their hard drives go away—then you've got nothing."[7]

Image, the third largest comic books publisher in the United States, should be applauded for allowing customers to store copies locally, but for all of the celebration of consumer ownership, its license terms aren't much different from those of other digital retailers. That license provides in part: "You shall not share, lend, lease, rent, sell, license, sublicense, transfer, network, reproduce, display, distribute, or otherwise make any Digital Comic available to any other person, to the extent that doing so requires making a copy of the Digital Comic (e.g., a copy on a hard drive, RAM, flash memory, a paper copy, etc.). A Digital Comic may be shared only by sharing the device containing the Digital Comic."[8]

So in Image's view, its customers "own" their digital comics, but are forbidden from doing most of the things we associate with ownership. They can't lend, give away, or resell a specific comic they "own" without also transferring their entire comic library and an expensive piece of hardware. That is an understanding of ownership many people wouldn't recognize.

The misuse of the language of ownership isn't limited to efforts to persuade the average consumer. HeinOnline hosts a massive database of legal publications, including law journals, judicial opinions, statutes, and treaties from around the world. Historically, it has offered these materials to libraries, law firms, and individuals on a subscription basis. Subscribers who sign into their HeinOnline accounts can access the wealth of information stored on its servers. In response to pressure from some subscribers, notably libraries, that value the comparative reliability, security, and privacy of local copies, HeinOnline launched its "Digital Ownership Program." Rather than signing up for a remote access subscription, by "purchasing digital ownership," users can "obtain ownership rights to PDF files" delivered on a hard drive.[9] HeinOnline does not provide a link to the terms of the Digital Ownership Program on its site. But after we asked for clarification, HeinOnline provided us with a copy. The relevant language is below:

> **V. PURCHASE TERMS**
> Customer may not: (i) sell, distribute, publicly display or in any other way exploit (commercially or otherwise) the Collection(s) or portions thereof, by any means, including, without limitation, sale, exchange, barter, transfer, assignment, or distribution, (ii) transfer, assign or sublicense any of the Customer's rights or obligations under this Agreement. ... Under the terms of this Agreement, Customer is authorized to make further copies of its original copy in perpetuity, as it may deem necessary, for purpose of preservation, refreshing, or migration, including migration to other formats so long as the purpose of such copying is solely for continued access to and/or archival retention of the Collection(s) in the manner permitted hereunder.

These terms make clear that HeinOnline's definition of "ownership" is an exceedingly narrow one. A library that purchases one of these hard drives couldn't lend it to another institution, for instance. In that sense, HeinOnline's understanding of ownership is even more cramped than the one offered by Image Comics.

Random House tried a similar ploy when its Vice President of Library and Academic Sales told *Library Journal*, "Random House's often repeated, and always consistent position is this: when libraries buy their RH, Inc. ebooks from authorized library wholesalers, it is our position that they own them."[10] We will address a range of issues facing libraries in the next chapter, but it is worth pausing for a moment to consider what it means for a library to own a digital copy. Most publishers refuse to deal directly with libraries when it comes to ebooks. Instead, they contract with vendors like OverDrive, who provide technology platforms that allow library patrons to access digital books. Given that baseline, it is easy to understand why the library community was puzzled by Random House's claim. When Peter Brantley of the University of California Davis Library sought clarification, Random House explained that by "ownership," it meant that libraries could migrate their ebook catalogs from one vendor to another. As Brantley put it, "That's very nice. It's just not ownership. It's licensing, with benefits."[11]

Contrast these claims of ownership with those made by technical publisher O'Reilly Media. When customers buy ebooks from O'Reilly they can "freely loan, re-sell or donate them, read them without being tracked, or move them to a new device without re-purchasing all of them,"[12] as long as they don't keep any copies of their books after lending or reselling them.[13] That's a notion of ownership that looks familiar to most of us.

Publishers and retailers understand the visceral appeal of the language of ownership. And they have succeeded in using that appeal to peddle digital products. But it remains to be seen whether we are actually getting what we bargained for.

The False and Deceptive Advertising Frameworks

Two distinct but overlapping bodies of federal law regulate the accuracy of claims used to market consumer products.[14] The Lanham Act—primarily the source of federal trademark protection—also prohibits the use of "any ... false or misleading representation of fact ... in commercial advertising or promotion [that] misrepresents the nature, characteristics, qualities, or geographic origin" of goods or services.[15] In addition, the Federal Trade Commission (FTC) is empowered by Congress to prevent the use of "unfair or deceptive acts or practices in or affecting commerce."[16] Both of these sources of law could be used to address potential mismatches between license terms and advertising claims. As an FTC official explained in 2009, "A company's marketing materials must be consistent with the nature of the product being offered. It's not enough to disclose the information only in a fine print of a lengthy online user agreement. ... If your advertising giveth and your EULA taketh away don't be surprised if the FTC comes calling."[17]

The standards for false and deceptive advertising under the Lanham and FTC Acts track each other fairly closely. We will discuss them in greater detail shortly. But first, we should highlight one important difference between these two legal regimes that relates to what lawyers call standing—the right to pursue your claim in court.

The Lanham Act creates a civil cause of action. That means a party who believes it was injured by false advertising can bring a suit in federal court against the advertiser. On its face, the statute creates broad standing. It allows "any person who believes that he or she is or is likely to be damaged by such act" to sue for damages.[18] Despite this inclusive language, courts have limited standing to competitors or others with a commercial interest implicated by the allegedly false statements.[19] Consumers, even though they are most directly harmed by false claims about the products they buy, are barred from challenging them under the Lanham Act.[20]

Courts—understandably concerned about opening the floodgates of litigation to every person upset that their toothpaste was not in fact "new and improved"—argue that competitors are in a better position to vindicate consumer interests than consumers themselves.[21] Competitors, these courts say, have greater resources and financial incentives to target false advertising. So we should expect them to vigorously pursue such claims.

Sometimes that is true, but not always. Companies make the expensive decision to litigate only if they think it will give them a competitive advantage. Suppose your competitor launches a highly successful, but arguably false, advertising campaign. You worry that you are losing sales because this competitor is overstating the benefits of their product. You could spend millions of dollars in legal fees in the hope that in a year or two a court will put a stop to the campaign and perhaps award you monetary damages. Or you could just adopt the same misleading tactic as your competitor. The incentive to adopt dubious advertising language grows as it becomes more widespread.

Of course, there are reasons to suspect individuals would be reluctant to challenge false advertising too. Aside from the most expensive purchases, the harm to a single person from a false ad is just too small to justify the time and expense of a court case. Class action lawsuits could solve that problem by bundling together the claims of similarly situated consumers in a single case. But without consumer standing, that option remains off the table as a matter of federal law. Consumers could sue under false and deceptive advertising statutes in various states. But those claims face their own set of hurdles. Because those laws vary in subtle but sometimes significant ways, class actions may be limited to consumers within a particular state. In addition, some retailers like Amazon include arbitration provisions in their terms that may well preclude litigation altogether.[22]

That's where the FTC's deception authority steps in. Section 5 of the FTC Act does not create a private right of action for consumers or competitors; it leaves enforcement entirely to the FTC.[23] Unlike competitors, the FTC is tasked with defending the public interest.[24] So we might expect it to serve as a more reliable proxy for consumer interests. But given its broad mandate and limited resources, the FTC has to carefully prioritize its enforcement efforts. So while it can't—and shouldn't—take on every questionable ad, the FTC can use its discretion to target deception that causes significant harm to buyers.

Whether it's a false advertising claim brought by a competitor or a deceptive practice case brought by the FTC, the central questions remain the same. In order to establish that an ad is false or deceptive, you must prove that it is misleading—that it is likely to convey a message that is not true. You must also prove that the message conveyed is material—that consumers would have behaved differently had they not been misled. So for example, a product advertised to U.S. customers as "Made in Turkmenistan" would be misleading if the product was actually produced in Tajikistan. But unless shoppers prefer products from Turkmenistan, that claim would be immaterial.

Misleading claims are not limited to express statements like "Made in Turkmenistan." They can be implied as well. An ad that claims a supplement will "boost your immune system" may not expressly promise to prevent the common cold, but it implies as much. Even omissions can be misleading. A television commercial for mail-order furniture, for example, would be misleading if it failed to disclose that the sofa depicted was child-sized. While express claims are clear on their face and require no additional proof of falsity, the meaning of implied claims and omissions are more ambiguous. In those cases, courts and the FTC consider other sorts of evidence including consumer testimony and surveys to decide whether an ad is likely to mislead.

How many consumers have to come away with an incorrect impression before an ad is considered false or deceptive? There is no precise answer to that question. Some courts talk about a "statistically significant"[25] or "not insubstantial"[26] portion of the intended audience. Others ask whether a "significant minority of consumers"[27] was misled. It's difficult to pin down an exact percentage, but prior cases suggest that a small number like 3 percent or 7.5 percent is below the legal threshold.[28] But courts have held that slightly higher rates like 10 , 15 , or 20 percent are enough to establish a likelihood of deception.[29]

The willingness of courts to accept survey evidence that demonstrates an ad deceives only a relatively small minority acknowledges that advertisements are often susceptible to more than one reasonable interpretation.[30] But where one of those interpretations is misleading, the advertiser is liable. It also reflects the fact that false advertising law is not intended to protect only the savvy or the skeptical. Its protections extend broadly to the public, "that vast multitude which includes the ignorant, the unthinking and the credulous."[31]

Once we know people are being misled, the question turns to whether or not those inaccuracies are material to their choices. Would they have behaved differently had they known the truth?[32] Perhaps they would have refused to buy the product, would have paid less for it, or would have preferred an alternative. Materiality can be presumed for express claims, implied claims intended by the seller, or claims that relate to the health and safety, central characteristics, purpose, performance, or cost of a product.[33] Of course people don't want products that are unsafe, don't perform as expected, or don't work for their intended purpose. For misleading claims that fall outside of these categories, courts and the FTC consider consumer testimony and surveys, among other evidence.

What "Buy Now" Means to Digital Consumers

Do the marketing efforts of digital retailers meet the definitions of false or deceptive advertising? To answer that question, Aaron Perzanowski and Chris Jay Hoofnagle conducted a first-of-its-kind study to gauge how consumers understand the phrase "Buy Now" as used by major digital retailers.[34]

This study surveyed nearly 1,300 likely buyers of digital books, music, and movies. The sample was representative of the U.S. population in terms of sex, age, and income.[35] Respondents were asked a series of questions about their media purchasing habits and, depending on their responses, were sorted into one of three groups: ebook shoppers, digital music shoppers, and digital movie shoppers. In order to better replicate real-world conditions, they were next presented with a number of popular media titles and asked to select the one that most appealed to them.

At this point respondents viewed mock webpages featuring the book, album, or movie they chose. An example is included in figure 5.1. MediaShop—the fictional online retail site respondents viewed—was created for the purposes of the study. Its design elements—the layout, buttons, product photos, and descriptions—would be familiar to any online shopper. Aside from its name, color scheme, and perhaps reduced visual clutter, MediaShop is indistinguishable from Amazon, iTunes, Target, or Walmart. Each respondent was presented one of four variations of the product page. One group saw the digital good they chose accompanied by a "Buy Now" button; a second group saw the digital good with a button that read "License Now"; a third group saw the digital good with a "short notice," to be described in

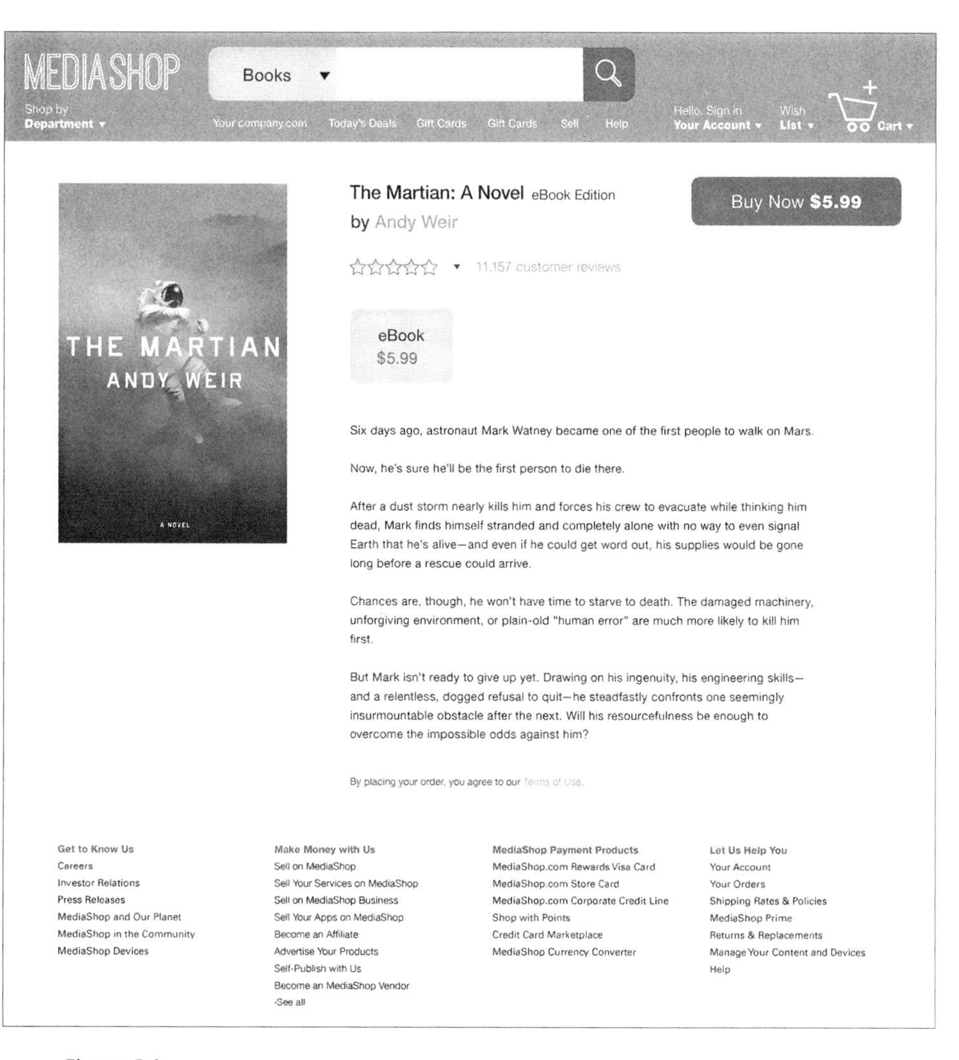

Figure 5.1
An example of a MediaShop product page

more detail; and the rest saw a physical good—a paperback, CD, or Blu-ray disc—with a "Buy Now" button.

Respondents were asked to review the MediaShop page as they normally would before making a purchase online. Notably, each digital product page included a link to MediaShop's Terms of Use.[36] Of the 956 respondents who viewed those pages, only 14 clicked the link to investigate the fine print. That figure—less than 1.5 percent—is consistent with other research that demonstrates how infrequently online consumers review terms.[37]

After completing their fictional transaction, respondents were presented a series of questions about what rights, if any, they obtained after paying for the product. As you can see in figure 5.2, lots of respondents believed that when they clicked "Buy Now" to acquire ebooks, MP3s, and digital movies, they were acquiring rights that we associate with ownership of physical goods. Survey respondents overwhelmingly believed that when they clicked "Buy Now" they owned the product that they purchased. Because ownership is a legal conclusion—one that is contested in the digital economy—it is hard to say with any certainty whether consumers are right or wrong when they express their beliefs about ownership. But on the whole, those beliefs seem to belie the claim often made by rights holders and retailers that people understand perfectly well that when they click "Buy Now" what they are buying is a license.[38] Putting aside the conceptual awkwardness of "buying a license," the survey data suggest that for a substantial number of consumers, the notion of buying entails a set of rights that are independent of any license terms.

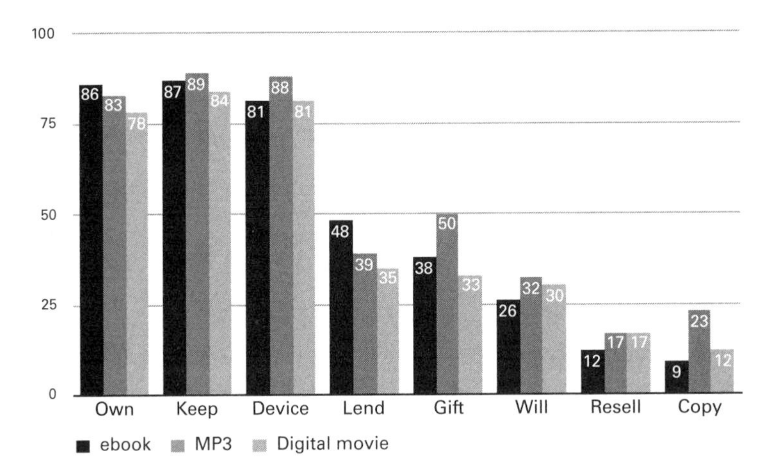

Figure 5.2
Percentage of respondents who believe the "Buy Now" button confers rights

For nearly nine out of ten respondents, "Buy Now" communicated that they were entitled to keep their digital purchase for as long as they wanted. That's typically the case with the physical media we buy. Once you pay for your hardcover book or your vinyl record, it is yours until you decide to part with it, barring theft or disaster. But as both a practical and legal matter, the same isn't true of the digital goods you buy. In chapter 1, we outlined some of the many ways buyers can be denied access to their digital purchases. Your retailer might go out of business or decide to shut down its servers as a cost-cutting measure. You might find your account wiped clean for violating the terms of service. You might wake up to find your device has been remotely deleted like the Amazon customers who thought they'd purchased *1984*. You might be denied access to ebooks you purchased months ago, like some Barnes and Noble customers, because your credit card recently expired.[39] Not only can your purchases be effectively repossessed after you pay for them, but risk-of-loss and termination provisions common in license agreements insulate retailers from any legal liability for denying you access to your purchases.[40]

An almost equally large majority of respondents believed that when they bought digital goods, they could enjoy them on the device of their choice, whether it be a laptop, smartphone, tablet, or dedicated reader or player. Assessing the accuracy of this belief is a challenge because of the variety of devices, formats, and business models in the marketplace. Some retailers have embraced the diversity of the digital ecosystem. Amazon, for example, supports a wide range of devices for digital media, from its own Kindle line to Apple iOS and Android devices, even including the latest Nook from competitor Barnes and Noble.[41] Amazon sees the ability to read ebooks on a buyer's device of choice as a selling point. Its choice to sell music in the de facto standard MP3 format paints a similar picture.

But other retailers have taken a decidedly more closed approach to device compatibility. Apple's iBooks can only be read on Apple devices. The same is true for iTunes music and movies. Through a combination of license terms, proprietary file formats, and DRM, Apple has inextricably tied the media it sells to its own hardware. In part, that's a reflection of Apple's longstanding obsession with carefully controlling end users' experience of its products. But it's also a function of the differing business philosophies of Apple and Amazon. Amazon works hard to keep prices low to attract an ever-larger customer base. It sells Kindle eReaders and tablets at break-even prices and may actually lose money on each sale.[42] But it hopes to profit in the long run by driving traffic to its site. Apple—despite selling billions of dollars' worth of apps, movies, and music—is in the hardware business.

And its profit margin on devices like the iPhone and iPad are as high as 69 percent, leading to quarterly profits of over $10 billion.[43] By ensuring that its customers can't play their media if they switch to a competing device, Apple keeps those profits coming.

Ultimately, whether buyers are correct in their belief about device compatibility depends on choices made by retailers, rather than their own legal rights. And in any case, that belief is often mistaken as a practical matter. In the MediaShop study, for example, the license limited respondents to the use of "Supported Devices." Only a handful knew that since the vast majority didn't read the license terms.

Lending is a widely recognized right of property owners. Book lending as a cultural practice predates the United States by several hundred years. And people have been lending music and movies as long as they have been available for sale. The same is true for gift giving. So it's hardly surprising that more than 40 percent of survey respondents believed they could lend and give away their digital purchases to friends and family. But it's standard practice for license terms to forbid such transfers. The Amazon Instant Video and MP3 stores, Apple iTunes, Google Play, Sony PlayStation Network, Microsoft Xbox Live, and countless smaller digital retailers explicitly bar consumers from lending, renting, giving away, or otherwise transferring their purchases.

Frustrated by the inability to make expected uses of their purchases, customers have pressured some retailers to liberalize their policies around lending and shared use. The Kindle and Nook stores both offer restricted lending programs. If publishers opt in, consumers can lend an ebook, one time only, for fourteen days. Of course, you can lend your hardcover books willy-nilly, whether the publisher likes it or not. Similarly, Apple's Family Sharing program allows digital media purchases to be shared among up to six accounts, provided they all share the same credit card information.[44] And while that might make it easier to ensure that the episode of *Peg + Cat* you bought on your laptop shows up on your kid's iPad, it's not the same as ownership.

Nearly 30 percent of respondents believed they could leave their ebooks, MP3s, and digital movies to loved ones in their wills. We've grown accustomed to inheriting physical media—a father's library or a grandmother's collection of LPs. And for many people, that tradition, or at least the expectation of it, survived the shift to digital copies. Locally stored copies can be transferred through a will easily enough: "I hereby leave my Kindle to my daughter." But tying a digital media collection to a single device is an impractical and incomplete solution. What happens when that device breaks? Or when the movie collection and music library on a single device is left to two people? And what about the cloud-based purchases that aren't stored locally at all?

Some early efforts to address these sorts of complications show promise. First, we've seen providers of web services developing tools to ease the transfer of accounts after the death of a user. Google's Inactive Account Manager and Facebook's Legacy Contact both allow users to designate a digital heir to take over their account should they meet an untimely end.[45] So far, digital media stores haven't rolled out similar tools. But it isn't hard to imagine them doing so in the future.

Second, lawmakers have taken some tentative steps to deal with the pressing problem of aging baby boomers with active online lives. Delaware became the first state to enact the Fiduciary Access to Digital Assets Act, a model law developed by the Uniform Law Commission.[46] That law gives heirs and other beneficiaries of an estate the power to control digital accounts and assets—including text, audio, video, and software—and to request transfers or copies of those assets.[47] This act, which has been introduced in a number of states, provides the first legal foothold for those seeking to control the disposition of their digital media collections posthumously. But the act contains a crucial limitation. Control over digital assets is limited "to the extent permitted under ... any end user license agreement."[48] In other words, if a license forbids this sort of transfer, your digital purchases die with you. For the time being at least, people who believe that "Buy Now" allows them to control their digital assets after death are mistaken.

Finally, we turn to the question of resale. Used booksellers have operated in the United States for centuries. Benjamin Franklin and Thomas Jefferson built their personal libraries in part by buying used books. And resale markets for records, CDs, videotapes, and DVDs—though not always embraced by copyright holders—have been fixtures of online and offline shopping for decades. But at this point, you don't need us to tell you that resale is nearly uniformly barred by digital retailer license agreements.

Perhaps because of the inherently commercial nature of resale, fewer respondents believed that "Buy Now" gives them a right to resell later. Of the questions surveyed, resale was the only one that did not result in a rate of deception well above the legal threshold. Nonetheless, the number who thought they acquired resale rights—12 percent for books and 17 percent for music and movies—makes a more than plausible case that a "not insubstantial" minority of consumers is likely to be misled.[49] Taken as a whole, this study suggests that when it comes to the rights we associate with ownership, the "Buy Now" button is a lie.

But it isn't enough to prove that the "Buy Now" button misleads consumers. The misinformation it communicates also has to be material to consumer purchasing decisions. If buyers wouldn't behave differently if they knew the truth, then they haven't been harmed by the deception. So the MediaShop study also gathered data on materiality. It did so in a few ways. First, it asked respondents to state their preferences when it came to the ability to lend, resell, and use their device of choice. Figure 5.3 illustrates the percentage of "Buy Now" respondents who strongly or somewhat preferred media purchases that allowed for those behaviors.[50]

There are two noteworthy findings here. First, a sizable portion of respondents—in many cases a majority—prefers to purchase goods that allow them to exercise rights we associate with ownership. Second, those preferences are remarkably stable across analog and digital goods. So consumer preferences for ebooks are very similar to their preferences for physical books. The same is true for MP3s and CDs, and for digital movies and Blu-ray discs. These rights remain just as important to buyers in the digital market as they have been in the physical one.

The survey also asked whether respondents would be willing to pay more for digital goods that they could lend, resell, or use on their device of

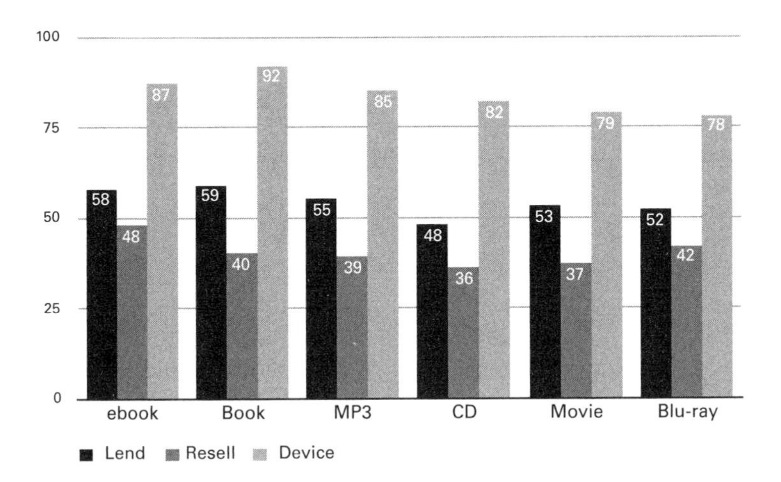

Figure 5.3
Percentage of respondents who express a strong or moderate preference for rights

choice. Most consumers were willing to pay more for at least one of those rights. The median price increase was $1, but the average was nearly $11 above the current Amazon prices. For the individual rights, respondents were willing to pay an average of $3.82 more for the right to lend, $3.24 for the right to resell, and $3.24 for the right to use media on their device of choice. Taken together, this evidence suggests that rights associated with personal property ownership influence the price of digital media goods. Roughly half of the respondents were willing to pay more for those rights. Since many respondents who expressed strong preferences for rights were unwilling to pay more for them, it is fair to conclude that some expect those rights to be part of the bargain under existing prices.[51] That information should interest retailers and rights holders.

Of equal interest, respondents were asked if they would be more likely to access digital books, music, and movies through subscription services—rather than purchasing them outright—if they couldn't lend, resell, or use their preferred device. Of the 94 percent of respondents who were familiar with subscription streaming services, more than half were more likely to stream if they could not lend their purchases. When it came to resale, 43 percent were more likely to stream. And 63 percent of respondents reported being more likely to stream if they couldn't use their purchases on their device of choice. For each of these rights, movie viewers were particularly susceptible to the draw of streaming services.

Fewer respondents—42 percent—had used or were familiar with the Pirate Bay and BitTorrent, two services associated with high volumes of infringing downloads. But among that group, 32 percent were more likely to download files without paying in the absence of a right to lend; 31 percent in the absence of a right to resell; and 40 percent in the absence of a right to use their device of choice. This suggests that the limited rights consumers acquire when they "Buy Now" contribute to the infringement of copyrighted works.

This survey evidence establishes that the "Buy Now" button misleads a considerable number of consumers about the legal rights they acquire when they spend money on digital goods. It also demonstrates that these misconceptions about their legal rights are material; people would behave differently if they knew that they didn't own their digital purchases. Applying the basic rules of false and deceptive advertising, the "Buy Now" button looks like an unlawful effort to exploit misinformation. The next question is: What can be done about it?

Coming Clean

The fact that a sizable number of consumers think they get a particular set of rights when they click the "Buy Now" button is not in itself an argument for granting them those rights. Consumer expectations are fickle things. They change over time and depending on context. They are manipulable. Rights—property, constitutional, or human—need a firmer foundation. In this chapter, we aren't arguing that the falsity of the "Buy Now" button helps us determine what rights people should have in digital purchases. Our goal here is more modest. It's to point out that the usual marketplace signals that tell us what consumers value have failed. They've failed because people incorrectly believe that they can do things with their digital goods that they can't. So the choice to embrace digital media does not prove, as some would conclude, that we've moved on as a society from the notions of ownership, lending, gifting, and reselling that have helped define our relationships with our property and each other for centuries.

No doubt, changes in the way we acquire, use, and share goods are underway. And they will have a profound effect on our culture. But those changes should take place in the open. Individuals should be fully aware so they can make thoughtful, deliberate choices. That only happens if they have accurate information. In some cases they do. Netflix and Spotify subscribers understand that once they stop paying their monthly fees, the movies in their queue and the music in their playlists go away. But we can't say the same for a la carte digital purchases.

That's where false and deceptive advertising law should come into play. There are two ways digital retailers could avoid deceiving their customers going forward. First, they could change the terms of their licenses to avoid misunderstandings that harm buyers. Instead of denying them economically valuable rights to lend, resell, and give away their purchases, licensors could grant them. Of course, retailers would need to negotiate with the copyright holders whose works they sell before making such a drastic change to their business models. Some retailers have taken tentative steps in this direction. We discuss licensed resale and lending solutions in greater detail in chapter 10, but for now we will just note that they strike us as both unlikely and problematic.

The other way to avoid deception is to change the way retailers talk about digital media transactions. If "Buy Now" fails to convey the limited set of rights defined by licenses, maybe we need a new button. Apple paid nearly $100 million to settle allegations made by the FTC that the company failed to adequately disclose that free apps targeted at children could be used to make in-app purchases. In response, Apple eliminated the "Free" button for those apps in favor of the less misleading "Get."[52] In theory, the same could be done with "Buy Now."

The MediaShop survey tested two alternatives to see if either could reduce consumer misinformation. As described earlier, some respondents were shown product pages that used the phrase "License Now" instead of "Buy Now." In theory, the word "license" would put consumers on alert that something was different about this transaction. But under the "License Now" formulation, respondents failed to consistently reduce misperceptions of their rights.

A more promising approach relies on what are called short notices. The theory behind a short notice is that if retailers disclose the salient facts about a transaction in a clear, simple way, people are more likely to understand that information. So even though we don't read licenses, short notices can not only inform us, but also help us make better decisions. From online privacy policies to HIPAA disclosures and credit solicitations, layered notice has been encouraged or required as a way to increase consumer comprehension of complex agreements or legal regimes.[53]

Another group of respondents in the MediaShop survey saw a digital product page that, instead of the "Buy Now" button, included a short notice that described their rights in clear, simple terms and using intuitive iconography. Examples are included in figure 5.4. For an ebook, for example, respondents saw a thumbs-up symbol informing them that they had the right to: download the ebook to approved devices, read the ebook

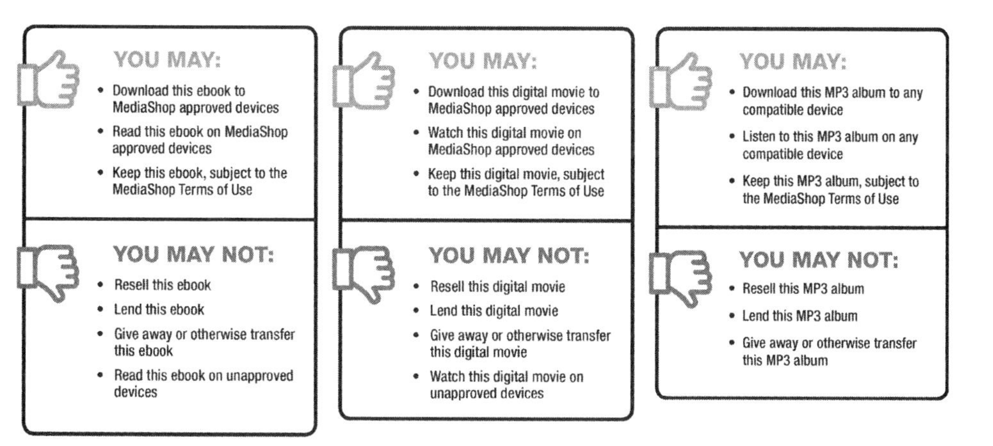

Figure 5.4
Examples of MediaShop short notices

on approved devices, and keep the ebook subject to the terms of use. A thumbs-down symbol was followed by text explaining that respondents did not have the rights to: resell the ebook, lend the ebook, give away or otherwise transfer the ebook, or read it on unapproved devices.

After viewing the short notice just one time, respondents were then asked the same questions about the rights they acquire after paying for a digital product. As figure 5.5 reveals, respondents demonstrated reduced rates of misconception under the short notice condition. Affirmative responses to the ownership question dropped significantly for all three media types. And yes answers to the lending and resale questions were cut by as much as half for ebooks and MP3s. When asked if they could leave their digital goods in their wills, ebook shoppers who saw the short notice were half as likely as their "Buy Now" counterparts to answer yes, a drop from 26 to 13 percent. Although outside of the range of statistical significance, MP3s saw an 11 percent drop. Likewise, when respondents were asked about the right to give away their digital media, there was a 10 percent drop for ebooks and a 14 percent decrease for MP3s.

The accuracy of respondents' beliefs was scored on a seven-point scale, with each correct response worth one point. Overall, the average score for all respondents was 3.1, with a median score of 3. Predictably, respondents who viewed physical media sold using the "Buy Now" button scored highest. Their mean score was 4.7 with a median of 5. Among respondents who shopped for digital goods, those who viewed the short notice performed

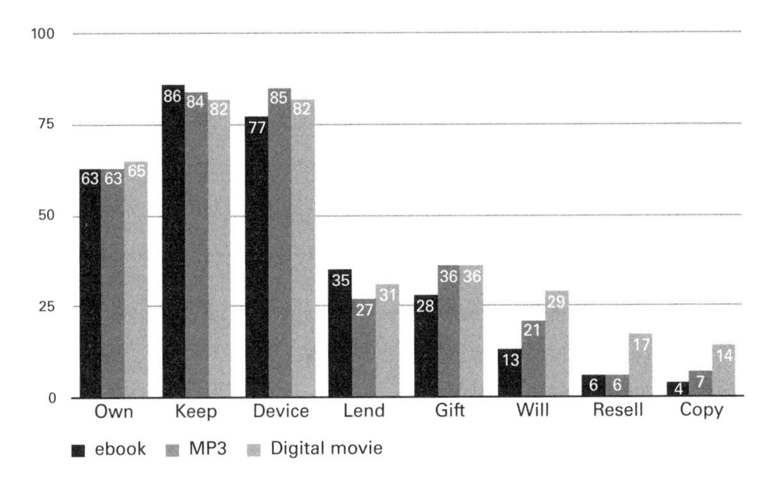

Figure 5.5
Percentage of respondents who believe the short notice confers rights

the best, with a mean of 3.0 and a median of 3. Those who viewed the "Buy Now" and "License Now" buttons scored considerably lower. The mean for "Buy Now" respondents was 2.45, with a median of 2. For "License Now" respondents, the mean was 2.27, with a median of 2. Compared to these two buttons, the short notice significantly improved how well respondents understood their rights after a single exposure.

Although the short notices were not perfectly or uniformly effective, they are a promising tool for preventing the kind of consumer deception that appears to be widespread in today's digital marketplace. Regulators should take a serious look at the efficacy of short notices and consider pressuring or requiring digital retailers to adopt them.

But more accurate disclosures are not a panacea. As valuable as accurate information about the nature of digital transactions may be, in chapter 6 we demonstrate that even highly sophisticated and informed digital media shoppers cannot avoid the constraints that law, licenses, and technology impose.

6 The Promise and Perils of Digital Libraries

In 1731, Benjamin Franklin and a group of his colleagues founded the Library Company of Philadelphia, what many believe to be the first public library in America and perhaps even the world.[1] Any member of the public could join the Company by buying "shares" that allowed one to use the library space and borrow any library book as often as it was available. Money from the sale of shares went toward the purchase of additional books for shareholders to enjoy.

Today, such a model of sharing is well-accepted practice. There are over nine thousand public libraries in the United States alone in addition to university and private libraries. For decades and in some cases centuries, these institutions have purchased books in order to allow their members and patrons to browse and borrow them. For many, the library-lending model is a hallmark achievement for education and public access to knowledge. Libraries function as archives of our cultural heritage, accessible spaces where communities gather and learn, and curators of specialized collections.

Legally and historically, the practice of library lending has depended heavily on the exhaustion principle embedded in personal property ownership. When a library buys a book, it exhausts the copyright owner's interest in that particular copy. The library can subsequently lend it out sequentially to any number of patrons for as long as it likes, even to other libraries through processes such as an interlibrary loan. It can also repair the book, make a small number of archival copies, and resell or donate the book at any time—all without needing to ask permission or pay the copyright owner additional money. Simply put, once the library buys the book, it owns the book, which allows it to distribute or dispose of that copy according to its own communal values, practices, and ethics—even if they diverge from those of the publishers. The same applies to videos, music, and most other forms of physical media that libraries acquire.

As we have noted, this model enables numerous benefits—privacy, simplicity, community, and discovery of new interests and areas of study. For example, librarians have for decades held to a strict ethical code that includes protecting patron privacy. We see this not only in the American Library Association code of ethics, but also enshrined in state laws, and in various political conflicts, where librarians have spoken out against government requests for patron records.[2] This level of commitment has served as a model, both ethically and legally, for other media privacy laws, such as the Video Privacy Protection Act and the California Reader Privacy Act. As Neil Richards notes in his book *Intellectual Privacy*, such protections are fundamental to both intellectual and academic freedom, among other democratic values.[3]

Even the notion of browsing information—something we now apply to websites or social media profiles—derives much of its cultural meaning from the way in which libraries have presented books in open stacks, free for all to peruse without prepayment, self-identification, or technological constraint.[4] Owning those books provides the basis for these freedoms and the institutional autonomy that libraries provide to their patrons.

Yet there is an undeniable tension between such property rights in physical media and intellectual property rights in the underlying works. Copyright owners have often cringed at the book-lending model, imagining that even under the sequential one-copy-per-patron constraints of analog media, libraries would cannibalize their sales if too many patrons could simply borrow a book, album, or movie instead of buying a copy for themselves.[5]

A fascinating example of this fearfulness appears in Ted Striphas's book *The Late Age of Print*. Striphas recounts how in 1931, a group of book publishers hired PR pioneer Edward Bernays—the "father of spin"—to fight against used "dollar books" and the general practice of book lending. Bernays decided to run a contest to "look for a pejorative word for the book borrower, the wretch who raised hell with book sales and deprived authors of earned royalties." The contest generated an impressive list of verbal assaults on those who would dare to lend or receive a book without paying for the privilege to do so. Suggested names included "book weevil," "greader," "libracide," "booklooter," "bookbum," "culture vulture," "bookbummer," "bookaneer," "biblioacquisiac," and "book buzzard," with the winning entry being "booksneak."[6] In the digital era, borrowing can be easier than ever. It doesn't even require a trip to your local library, if you can check out books from your e-reader. So the idea of frictionless digital book lending has some publishers absolutely terrified.[7]

Yet there is little doubt that digital lending and ebooks are critical to the future of libraries. Every year, ebook acquisitions continue to rise. For example, from 2010 to 2011, academic libraries increased their total ebook holdings from 158.7 million to 252.6 million.[8] In 2012, the American Library Association reported that 76 percent of public libraries offered free access to ebooks to library patrons—up over 20 percent since 2009.[9] A recent PricewaterhouseCoopers study suggests that the percentage of ebooks sold in the United States and Great Britain will surpass that of print media (including audiobooks) by 2018.[10] Library spending on children's, juvenile, and young adult ebooks in 2014 grew by 48 percent over 2013.[11] And according to the CEO of OverDrive, Inc., one of the dominant U.S. ebook providers, "Ninety-three percent of children between the ages of two and thirteen are reading or being read ebooks at least once per week."[12] While analog books don't appear to be disappearing anytime soon, ebooks are quickly becoming a centerpiece of what patrons want from their library's digital collection.[13]

So why does this matter? Won't libraries simply lend ebooks the same way they lend physical books now? Unfortunately, the answer is unclear because of differences in the distribution schema for analog and digital. First, as we've already discussed, borrowing digital books can result in the creation of additional copies on the computers, phones, or other devices patrons use to read them.[14] These extra copies arguably infringe on the copyright owner's exclusive right of reproduction, unless they fall under an exception or limitation, such as the exhaustion principle or the fair use doctrine. Second, because ebook sales are largely modeled on software sales, they often come with complex licenses that muddy the waters around ownership. Since libraries don't "own" the ebooks they buy in the same way as their physical book holdings, they can't rely on the simple rules of exhaustion to actuate large-scale lending on their own terms. While a few ebook publishers have allowed libraries to retain traditional ownership rights in ebooks, most publisher ebook licenses now attempt to dictate the precise terms under which libraries make works available to their patrons.

If this world view holds, then the shift to ebooks will change many fundamental functions within libraries—from acquisition and lending to archiving and fundraising. And it will have a profound effect on the benefits that ownership and exhaustion have historically provided, including privacy, simplicity, preservation, and community.

Fabricating Friction

Most libraries believe in broad public access to their holdings. However, with ebooks, the introduction of licensing models rather than sales has complicated, and some would say undermined, the library's mission. On the one hand, no one disputes that access to ebooks increases access to cultural heritage and scientific knowledge. On the other hand, ebook licenses often incorporate artificial restrictions. Publishers may insist on these provisions in order to introduce artificial friction between libraries and their patrons, to keep readers from becoming digital "booksneaks" and using libraries as a substitute for purchasing traditional and digital books.

There is no shortage of examples of this artificial friction. Publishers often limit the availability of titles by withholding them from circulation throughout a given year. They impose distribution delays by enforcing waiting periods between patron loan requests and downloads. They restrict lending geographically by deciding where a customer can borrow a book and even where they can read it. They cap the number of books each patron and each library can borrow and lend. And they charge libraries based on the number of times a book is lent instead of on a per-title basis. None of these limitations on libraries and their patrons exist for analog books. Library ownership of the books exhausts any attempt by publishers to assert such control. Yet ebook publishers use licensing and other technological constraints to attempt to wrestle back control over the world of digital library lending. And while there is certainly an appeal for libraries, who often suffer from severely constrained budgets, to embrace a more "on demand" and single-serving book acquisition business model, these practices when taken together raise real questions about the long-term impact they will have on library collections.

At first blush, such artificial friction may seem like an equitable balancing of intellectual property rights with digital media ownership. Even though such friction is unenforceable as a matter of copyright law for analog media because of exhaustion, those inefficiencies do attempt to simulate various market effects that would, in theory, reduce the impact of library lending of ebooks on publisher sales. The more friction patrons encounter, the more likely they will pay for the ebook instead of borrowing it from their local library for free.

Yet much like artificial gravity, there is a sense that such systems are a cheat of sorts. Rather than adapting to the new digital environment, these tactics seek to imperfectly impose restraints that would not naturally exist but for copyright holder concerns. Why shouldn't public libraries struggling under ever-increasing financial constraints be allowed to capitalize on the benefits of digital copying, especially when publishers benefit from the decreased costs of digital production and distribution.

From a purely economic perspective, artificial friction may well alleviate the concerns of the media industries and even save libraries money if the pricing is fair. Yet copyright law was never designed solely to benefit private market actors. Instead, as a constitutional matter, U.S. copyright law was intended to use private market incentives in ways that ultimately benefitted the public at large, not exclusively or even primarily copyright holders. Thus, as enamored as some of the founders might have been with the romantic ideal of authors and inventors, it was ultimately public access to knowledge and the resulting "progress of science and the useful arts" that was the true metric of IP's success.

A library's ownership of its media—books, music, movies, newspapers, photographs, or software—vastly increases public access in ways that the private market alone cannot.[15] This is true for both analog and digital media. The more friction one puts between the public and library holdings, the fewer patrons have access to those holdings.

Of course, copyright owners argue that unless they profit sufficiently, they won't invest in the production of new works, which would result in the public having nothing to access. This well may be true at some point, but the exhaustion principle guarantees copyright owners at least one purchase per copy already—thus fulfilling some part of copyright's bargain between the public and the author. But what if this isn't enough in the digital age? As the Copyright Office asserted in a special report on "digital first sale" in 2001, "the potential harm to the market and increased risk of infringement that would result from [a digital exhaustion rule] could substantially reduce the incentive to create."[16] While it is true that digital copies lack both the friction of physical ones—in other words, the time and energy it takes to transfer a copy from one person or institution to another—and the same decay rate, there is still no question that initial digital sales are providing substantial compensation to copyright owners with significantly reduced costs for production, distribution, and inventory. Allowing transfers of rivalrous digital rights, consistent with the exhaustion principle, could provide much of the friction of physical books. The key is finding systems that continue to balance these objectives for digital works in the way that exhaustion has done historically.

Moreover, copyright has always coexisted with individuals and institutions owning copies as personal property. The idea that personal property rights in copies should always be subservient to copyright interests presents a dangerous precedent for property rights in general. A shift from balancing copyright ownership with library media ownership to one where control is entirely within the hands and licensing terms of copyright owners raises great concerns.

Libraries without Collections

Let's take a step back and think about what all of these changes might do to the relationship between libraries and their patrons. As the inscription on the Boston Public Library facade proclaims, books on the shelves within are "free to all"—not only in the sense that no payment is required, but also in the sense that they come without strings attached—free from restraint, obligation, and complexity. However, in a world where every publisher insists on a different set of license terms and every ebook platform or DRM provider layers their own business models, software, and implementation on top of those licenses, library patrons nowadays are anything but free from complexity and restraint.[17] That complex patchwork has created real problems for libraries and their patrons. One recent study found patrons suffered through an average of nineteen clicks in order to check out a single ebook from most public libraries.[18]

Libraries have responded to this in several ways. In order to act as a buffer for patrons, many have tried to shoulder the burden by negotiating licensing deals with ebook vendors and publishers. This has led to serious dependencies. For example, most people have never heard of OverDrive, Inc., but this Cleveland, Ohio-based software vendor services over 90 percent of the library ebook market.[19] Other vendors include 3M and Baker & Taylor.

These vendors provide an electronic gateway that connects publishers to libraries and their patrons. They allow libraries to license ebooks stored on vendor servers, and using vendor software the ebooks are transferred to patrons for temporary use on their phones, tablets, or computers. At first, this seems innocuous enough and perhaps even ideal, as libraries can simply defer all customer service and technical issues to the vendors directly. However, this shift in the architecture of ownership and power creates an entirely different dynamic among publishers, vendors, libraries, and patrons. Prior to these systems, libraries would simply buy books from wholesale vendors, or occasionally directly from publishers, and maintain full control over their offerings. Library staff decided how to organize the books on the shelves, how long to allow them to be lent out, and what records to keep about their usage. Under the exhaustion rule, once the library purchased a copy of a book, the publisher and the distributor have absolutely nothing to say about how, when, to whom, or how often that copy was lent to a patron or institution. Now, even the most prestigious libraries are often beholden to intermediaries such as OverDrive for many of these functions. Upstream providers control when and how books are available, which titles persist and which disappear from digital shelves and search queries, and often, which patrons may or may not access them and under which circumstances.[20]

For example, in 2011, HarperCollins, a major publisher, announced that it would only allow libraries to lend its ebooks twenty-six times before forcing them to expire. HarperCollins claimed these self-destructing books were calculated to represent the rate of physical decay in analog copies.[21] If a book is lent to patrons for two weeks at a time, that means HarperCollins expects libraries to replace popular hardcovers every year. Regardless of the accuracy of that estimate, this shift—from lending and borrowing as a normative, communal practice governed by copy ownership and internal library policies to a model that allows publishers to define the legal terms and technological conditions under which libraries lend books—raises serious concerns.[22]

In other instances, works simply aren't available on a platform or in a medium that allows for lending. Kevin Smith, director of Copyright and Scholarly Communications at Duke University Libraries, has documented the dearth of options facing libraries in one such case. He described a new recording of celebrated conductor Gustavo Dudamel and the Los Angeles Philharmonic that is only available as a digital download via iTunes. As Smith explains, "The licensing terms that accompany the 'purchase'—it is really just a license—restrict the user to personal uses. Most librarians believe that this rules out traditional library functions" like lending.[23] When librarians tracked down Universal, the copyright holder in the recording, it offered to provide them an educational license for use of 25 percent of the album. That license would last only two years and would run them $250 in processing fees plus an unspecified additional amount. That's the tangled web of licensing and negotiation with which libraries must now contend. Just a decade ago, a library could have bought the entire recording on a CD for less than $20 and lent it as it pleased.

Fortunately, it appears that some publishers are responding to these concerns with more progressive policies. For example, Penguin Random House (now consolidated after a merger) will be offering its adult and children's frontlist and backlist digital titles under a "one-e-Book and one-user" policy and dispensing with its one-year lending cap on all ebooks. Libraries will now be able to loan the book out to as many patrons as they want as long as they follow an "exhaustion-like" single copy per patron rule.[24] Skip Dye, vice president of library sales for Penguin Random House, described the revised policy as an "opportunity for the full and permanent ownership of our titles purchased for [library] collections, which can evolve into a potentially unlimited number of library patrons borrowing that e-Book in perpetuity." This is a significant win for the library community; however, it is worth noting that it took nearly five years to negotiate the terms back to their analog equivalent.

Another example of expansive vendor control over library ebook lending is the use of proprietary software to define how patrons access ebooks from their phones, tablets, computers, or other devices and how libraries facilitate that access. Most of the time when a patron selects a book to check out from the library's catalog, they do so through the library website or app. However, as soon as the patron selects which book to read, the library is left out of the loop. The vendor takes over, transferring the ebook to the patron, and governing their interaction with the content.

At first, this may seem like just another technological evolution. However, it has serious implications for libraries and their patrons. For example, some vendors reserve in their software terms a unilateral right to terminate ebook access of any patron or library in the event that the vendor determines, in its sole discretion, that a patron or library fails to comply with the vendor's terms and procedures. In other words, if the vendor decides its terms have been violated, it can cut off a community's access to its ebooks. Imagine if Random House could walk into any library in the country and pull all of the books it published from the shelves if it suspected that a patron had made some objectionable use of them. That is the power that these vendors are now claiming.

Libraries have responded in a variety of ways, both institutionally and technologically. The ALA, its members, and other library associations have stepped up their emphasis on negotiating greater control in vendor contracts, particularly around patron privacy, a topic to which we will return. In addition, a new project called Library Simplified, a joint effort of public libraries in Boston, Cincinnati, New York, and Sacramento, among others, is seeking to create a special ebook reader—one made for libraries, by libraries—that consolidates and automates this complicated set of interactions, and reduces the number of clicks required to check out an ebook from nineteen to three. By reestablishing control over their relationship with their patrons, libraries may gain back some of their historical control over access to knowledge and patron data, which they often protect more vigorously than commercial vendors do.

Libraries and Cultural Preservation

As we have noted, libraries, museums, and archives all serve an important function in preserving our culture, our history, and various forms of knowledge. Yet these functions depend inherently on these institutions having control over the works they acquire. Traditionally, control came concurrently with book ownership, as acquisition of the physical property rights in books provided libraries with the authority they needed to decide how, when, and where, and by whom it would be held. It also exhausted intellectual property rights in the book, preventing any interference with the library's mission by copyright owners. With ebooks, this mission is much more complex and challenging to fulfill. On the one hand, digital books are easier to store—they take up less physical space and can be moved more easily. But as we've noted, publisher- or vendor-imposed licenses and technological restrictions on ebooks introduce new problems.

These problems become especially acute for works at risk when their economic value may be less than their cultural value. In such situations, libraries as well as other participants in secondary markets, such as used bookstores, have greater incentives than publishers or ebook vendors to maintain copies of books since publishers can charge a premium on newer versions. First editions or recent textbooks are good examples. For analog books, this discrepancy between profit and preservation objectives can lead to situations where institutions such as libraries are willing to pay to purchase or digitize older works, but the works' copyright owners have either gone out of business, disappeared, or become impractical to find.

For these "orphan works," libraries often preserve physical copies and, in limited circumstances, make digital ones available to patrons. And while there is some concern that copyright owners might come out of the shadows and reclaim their orphaned works, there is a strong case that such forms of digital preservation and access qualify as fair use, in part because many libraries map digital access to physical holdings on a one-copy-to-one-copy basis.[25] In a world where copies reside on publisher or vendor servers, subject to restrictive license terms, the virtual holdings of every library are at risk of vanishing, especially if they are orphaned.[26] This fear has already become reality in the digital music industry, raising concerns at the Federal Trade Commission.[27]

Consider a recent preservation project at Yale University to archive 2,700 VHS tapes from the 1970s and 1980s featuring so-called "Scream Queens," horror and exploitation movies emblematic of "the home-video revolution of the time, as well as the cultural mores and politics of the Reagan era they emerged in."[28] While many might consider such a collection uncouth or bizarre, to cultural critics it "tell[s] the story of a particularly significant gap between the old Hollywood model of the '50s and '60s and the corporate mergers of the '80s that created today's modern media behemoths. In the era of video tapes, independent producers and distributors could reach a mass audience using cheap technology and local stores, both of which lowered the profit threshold for moviemakers."[29] Harvard and Cornell have taken on similar efforts to collect archives related to the emergence of hip-hop, and New York University has acquired its own collection of cultural artifacts related to the rise of Riot Grrrl, an underground feminist punk movement in the early 1990s. Such preservation efforts mainly come from secondary collectors, not the original publishers. In fact, in counterculture or low-budget genres such as these, publishers often go in and out of business quickly and are nearly impossible to track down in order to secure various legal permissions. Were these collections held in digital form on now-defunct vendor servers or controlled with proprietary vendor technology, it might have been impossible to save them for historical, cultural, and educational purposes.

The commitment to preservation itself is also cultural. Ownership of works over their lifetime promotes long-term thinking. As works age, librarians, archivists, and museum workers are continually reminded of their duties to retain these objects in ways that do not diminish access. Ephemeral "on demand" access systems, intangible licensed rights, and technological control mechanisms discourage these approaches and instead focus on more short-term goals such as convenience and instant gratification. Not that these short-term goals are unimportant or undesirable. In fact, they are some of the great benefits of the digital age. Librarians have been among the best at recognizing these benefits while at the same time understanding the long-term challenges.

In response, cultural institutions including many libraries are working to establish digital means of ensuring preservation. Efforts such as the Digital Preservation Network and Academic Preservation Trust are working to build federated "dark archives" that will keep redundant copies in case of catastrophic loss of originals. In order to do this, these efforts depend on both the doctrine of fair use and, in some cases, the narrow preservation provision in the Copyright Act to shore up the gap between what exhaustion previously provided and where digital libraries and archives sit today.

When Copyright Owners Attack: IP as an Adversary of Preservation

Lack of perceived profitability isn't the only problem for preservationists. Modern history is replete with cases involving efforts to limit or decimate library holdings, often by political groups or governments.[30] While most of these challenges have been via political muscle, copyright holders have also sought to censor access to works. Most famously, the German government, copyright holder of Adolf Hitler's *Mein Kampf*, has prohibited the book's publication in Germany for decades, and only now must allow it for the first time in seventy-five years because the copyright has finally expired.[31] Here in the United States, we have seen similar attempts to use copyright law for purposes that work counter to the goals of preservation.

Take, for example, the case of *Worldwide Church of God v. Philadelphia Church of God, Inc.*[32] WCG was founded in 1934 as the "Radio Church of God" by Herbert Armstrong. Armstrong held the title of "Pastor General with the spiritual rank of Apostle" and led the church until his death in 1986. Along with many other publications, he wrote a 380-page book entitled *Mystery of the Ages* (*MOA*), of which WCG distributed over nine million free copies.

After Armstrong's death, WCG decided to stop publishing and using *MOA* for several reasons, including the fact that the church's positions on various doctrines such as divorce, remarriage, and divine healing had changed. Philadelphia Church of God (PCG), a rival whose members claimed to follow the "authentic" teachings of Armstrong, seized this opportunity and began printing and distributing *MOA* in its entirety. WCG sued PCG for copyright infringement and won, halting publication not because WCG would lose profits, but because WCG did not want PCG patrons to read it.

Now, to be fair, WCG did not request that anyone who already had *MOA* rid themselves of their copies or that any public libraries or archives destroy copies they owned. But it is important to note that WCG also lacked the legal authority to demand such actions. Copies of *MOA*, even infringing ones, cannot be reclaimed once sold because the copy is owned as personal property by the purchaser.[33] However, for digital copies that libraries don't own, any copyright holder who wants to remove a book from the shelf could simply terminate the libraries' license and remove the book.

A more recent example involved the best-selling book *The Boy Who Came Back from Heaven*, allegedly recounting the story of six-year-old Alex

Malarkey's visit to heaven after being injured in a car crash. Nearly five years after publication, Malarkey admitted that the story was fabricated, prompting its publisher to take "the book and all ancillary products out of print."[34] While some were sympathetic to the desire to withdraw the book, others saw it as an important flashpoint in an ongoing cultural and political dialogue about religious communities in America, part of a popular genre of "heavenly tourism."[35] Because of exhaustion, all analog copies of the book are still available to be preserved, analyzed, assigned in classes, and critiqued over any objection from the authors or publishers. The fate of the digital editions is less clear. As Amazon demonstrated with the remote deletion of *1984*, there is real risk of disappearing titles when copyright owners object to their existence.

Libraries and Safeguarding Patron Privacy

Libraries have also historically been safe spaces for readers who wish to protect their privacy.[36] This is not only due to the strong legal and ethical codes protecting library records from disclosure, but also the physical ownership of library media. Once the library purchases a work, the copyright owner has no legal interest in that particular copy anymore and cannot track or meter its use or whereabouts. Contrast this with ebooks that libraries must license. Even in the hands of libraries and their patrons, publishers can use a combination of license terms and technological controls to track their use. This raises a host of privacy issues, including potential chilling effects on those who would seek out controversial or revealing subjects such as medical treatments, sexuality, or unpopular belief systems.[37]

Moreover, the danger to patron privacy becomes amplified in a system where multiple parties have an interest in and access to the ebook distribution chain and related patron data. When a library owns a book, it can decide what patron records to keep and who can view them. Most ebook providers require that readers share data with multiple vendors, from DRM suppliers and e-reader app makers to the original publisher. Vendors may keep records on every transaction that flows through their servers, including which books you've checked out and which you've placed on hold for future reading.[38] Adobe's software has even tracked each page you've read and how long you lingered on it. Some emerging library standards are moving toward demanding strong privacy protection from ebook vendors, but such protection is no longer a given in a world where libraries must negotiate for it instead of one where they own and control the books directly.[39]

Our constitutional right to privacy that protects records of our reading habits from government surveillance and law enforcement subpoenas also depends, in part, on property rights in the media we access. The Fourth Amendment protects "the right of the people to be secure in their persons, houses, papers, and effects." Our "papers" include the things we write and the things we read. And our "effects" include the property we own. The Fourth Amendment was intended as a buffer between what we read and write and the government's interest in gathering data on its citizens.[40] Obviously, we don't own the books we borrow from the library. But through both state and federal statutes as well as keystone court decisions, it is well established that libraries can object to inappropriate government requests for library records on our behalf.[41] But when that information is stored as part of a commercial transaction with vendors and publishers, it is often no longer within the protective ambit of the library's code of ethics or statutory protection. Instead, it potentially falls within what's called the third party doctrine, which holds that once a consumer voluntarily shares information—like what books they read and when—with a commercial entity, they may no longer have a reasonable expectation of privacy in that information, and the protections of the Fourth Amendment may no longer protect that information from disclosure.[42]

Yet why should we care if the government accesses the records of what we read or watch? Intellectual privacy of this sort is fundamental to a functioning democracy. As Justice William O. Douglas observed, "Once the government can demand of a publisher the names of the purchasers of his publications ... fear of criticism goes with every person into the bookstall ... [and] inquiry will be discouraged."[43] The most blatant example of such criticism and the anti-democratic effect it can have arose during the anti-communist witch hunts of the 1950s and 1960s. At the McCarthy hearings, many of those called to testify were questioned on whether they had read Marx and Lenin.[44] They were asked whether their spouses or associates had books by or about Stalin and Lenin on their bookshelves.[45] Congress even passed a law requiring individuals to file written requests with the U.S. Postal Service to receive "communist political propaganda" through the mails until the Supreme Court struck it down because it was "almost certain to have a deterrent effect" on speech and association protected by the First Amendment. The Court especially noted that "public officials, like schoolteachers who have no tenure, might think they would invite disaster if they read what the Federal Government says contains the seeds of treason."[46]

This threat is not merely hypothetical. There have been several famous cases of government agents seeking lists of what we read and watch. One of the most prominent involved Monica Lewinsky, the White House intern involved with President Bill Clinton. In his investigation as Special Counsel, Kenneth Starr issued subpoenas to Barnes & Noble and Kramerbooks, an independent book store in Washington, D.C., for a list of all Lewinsky's purchases over a thirty-month period. Kramerbooks fought back and went to court to protest the subpoena, asserting that the First Amendment protected readers from the chilling effect of the government knowing what they were reading.[47] Eventually, Lewinsky's lawyers turned over some of the information directly to Starr, and the bookstore was never required to comply with the government's request.[48]

Nor are such witch hunts solely vestiges of the analog era. In 2007, federal law enforcement came knocking on the door of Amazon.com, asking for the reading records of 120 of its customers. Amazon fought back, successfully convincing the trial court to reject the subpoena. In holding so, the court wrote: "If word were to spread over the Net—and it would—that the FBI and the IRS had demanded and received Amazon's list of customers and their personal purchases, the chilling effect on expressive e-commerce would frost keyboards across America ... well-founded or not, rumors of an Orwellian federal criminal investigation into the reading habits of Amazon's customers could frighten countless potential customers into canceling planned online book purchases, now and perhaps forever."[49]

Studies have confirmed this chilling effect. One survey found that 8.4 percent of Muslim Americans changed their Internet usage because they believed their habits were being tracked by the government.[50] Even the controversial section 215 of the USA PATRIOT Act, which the National Security Agency used to justify collecting millions of American phone records, was originally envisioned as the "library provision" that would allow the U.S. government to demand any patron's library records simply because they were somehow relevant to a terrorism investigation.[51]

It is reassuring that both commercial book vendors and libraries have stood up for the privacy of information about our reading habits, and perhaps they will be able to continue to do so even in the age of the ebook.[52] But what if they don't? Do we have any rights to stop them from turning over our information? The further up the chain the information travels, the less claim we have to privacy. It's one thing for your local library or bookstore to assert itself as a custodian of the record of your purchases and stand in your shoes to fight for your privacy; it's another to claim that your book-viewing data, passed from device to provider to publisher, is somehow still yours. Thus, cloud storage and streaming books may also shift our sense of intellectual privacy if we are not able to secure it. Fortunately, California has taken a strong step in this direction by passing the Reader Privacy Act, which requires all vendors of electronic books and online book services to respect patron privacy. Perhaps other states will do the same.

Libraries and Innovation

In 1894, Historian John Willis Clark gave a lecture at Cambridge University entitled *Libraries in the Medieval and Renaissance Periods* in which he stated "[a] library may be considered from two very different points of view: as a workshop, or as a Museum. ... Mechanical ingenuity ... should be employed in making the acquisition of knowledge less cumbrous and less tedious; that as we travel by steam, so we should also read by steam, and be helped in our studies by the varied resources of modern invention."[53]

How does one "read by steam" in the digital age? Numerous library-related entities are exploring that question, from the Internet Archive's Open Library to the Digital Public Library of America.[54] Even the New York Public Library has a geek team, a group they call NYPL Labs.[55] NYPL Labs has produced many interesting projects to date—from annotating Google Maps of New York City with photos from their city archives to assisting scientists in analyzing climate change by tracking fish prices from nearly a century of digitized New York restaurant menus. All of this is possible because they own the physical materials and thus, digitizing them for analysis is a much simpler project. Consider, however, materials that are licensed and not owned. How does a library expand the public's understanding and engagement with materials when they belong to someone else and sit on remote servers they cannot access?

Or consider HathiTrust, a consortium of digital library efforts.[56] HathiTrust houses well over five million digitized books, the vast majority of which were scanned on behalf of the libraries by Google. When the Authors Guild sued HathiTrust for copyright infringement of these books, it asserted that the libraries had no right to lend the physical books they owned to Google for scanning purposes or to use the digital copies that Google provided them in exchange. Yet the courts that ruled on the case held that these actions were fair uses. HathiTrust transformed these paper-and-ink books into a massive digital archive and database, an altogether different sort of work, suitable for very different purposes. At the same time, it greatly increased access to knowledge.

What would have happened if those books had not been physically on the shelves for the library to lend to Google, but rather on the servers of OverDrive or various publishers? If the libraries tried to hand over to Google millions of ebooks, publishers and vendors would have pointed out that independent of copyright concerns, this violated the terms of their license agreements. The fact that access was conditioned on a license rather than ownership also could have changed the fair use analysis significantly. One fair use factor courts consider is the impact of the use on the market for the copyrighted work.[57] And if libraries are already negotiating and agreeing to license terms, presumably they could have paid more for terms that contemplated these sorts of uses. In this hypothetical scenario, the fact that libraries neglected to acquire such rights could be interpreted—wrongly, we think—by some courts as weighing against fair use. Regardless of the outcome, ownership of the books gave the libraries a certain independence from publishers as a practical matter. A library without physical books would be faced with the risk of losing access to their entire collection of ebooks if their actions upset vendors and publishers, putting them in a precarious position to fight for fair use and academic freedom in the first place. The security of owning the physical copies of the books provided libraries with the strength to stand up for what was ultimately ruled to be legal.

A Library with No Friends

Scattered across the United States are countless "Friends of the Library" groups. These supporters exist to raise money and help local libraries thrive in their communities. One of the main ways they do this is to host book donation efforts. These efforts ask local citizens and institutions to donate old books—not for the libraries' shelves, but to resell to help raise money for new library purchases. It is one of the most time-honored ways to give back, by giving away your old books so that the library can turn them into new ones.

But every single aspect of such fundraising depends on ownership. If patrons who buy ebooks don't own them and libraries they support can't own them, then how does one donate an ebook to one's local library at all?

This is not just a problem for libraries, but for many other access points for knowledge and cultural heritage. For example, consider Project Cicero,[58] an annual nonprofit book drive designed to create and supplement classroom libraries in under-resourced New York City public schools. Since 2001, Project Cicero has distributed 2.3 million books to more than 13,000 New York City classrooms, reaching over 550,000 students. It receives new and used book donations from more than one hundred independent, public, and parochial schools each year.

But what will the future of such projects look like in a world when parents, teachers, students, and schools no longer own the books they use and read? What happens when your Kindle or iPhone won't let you donate your book? Or the terms of service for your ebook provider or the license agreement on the book itself forbid it? Or copyright law deems you an infringer for donating a used ebook to your local public library?

The problems facing libraries are, in many ways, the same problems confronting consumers writ large. Complex license terms, uncooperative technology, and outdated copyright laws interfere with the kinds of uses they've made for centuries. While some might suspect our fellow citizens of uncertain motives or questionable intentions, by focusing on libraries—a set of institutions and actors with a well-deserved reputation as responsible actors—it is easier to understand that a digital exhaustion doctrine is not meant to provide refuge for scofflaws and infringers. Instead, it's a way to protect the network of socially valuable uses that owning books makes possible.

7 DRM and the Secret War inside Your Devices

In Ray Bradbury's iconic dystopian novel *Fahrenheit 451*, a war rages in a future society over the existence of books. Those in power seek to destroy them, both because of the controversial ideas they disclose and because of their perceived limited utility in a society filled with video-enabled walls and mobile media devices. Those who rebel against these rules hide books to preserve them for historical, political, and philosophical reasons. Bradbury's infamous firemen—shock troopers who kick down doors and incinerate homes where books are hidden—and their mechanical drone-like hounds that sniff out literary contraband are meant as provocations to incite our fears that the very book we hold in our hands might be taken away from us at a moment's notice in the name of "public happiness." By personifying this version of absolute control, Bradbury makes clear that notions of personal property or domestic privacy stand no chance in a society that values centralized authority over individual autonomy and cultural heritage.

As a commentary on the McCarthy Era, Bradbury's work is a reaction to a specific threat to our engagement with ideas and the cultural artifacts containing them. And although the particular brand of control Bradbury had in mind has not manifested itself in contemporary U.S. culture, there is a different sort of threat to our freedom to read, explore, and share ideas—one that is more subtle, but all the more dangerous for it. This threat doesn't kick down your door in the dead of the night; it already lives in your home. It's embedded into the media you buy and stored on the devices you carry in your pocket. It doesn't rely on physical force or the power of the state to enforce its rules, just the often unseen operation of software code.

Digital Rights Management (DRM) is the euphemism for a range of technologies implemented by copyright holders, device makers, retailers, and other intermediaries designed to control how, where, when, and whether consumers can use their books, movies, music, and other content. In a nutshell, DRM is a digital guard capable of silently monitoring your digital activity and enforcing any restrictions or limitations demanded by rights holders. DRM can prevent you from copying a file, even for legally permissible reasons like personal backups. It can restrict your iTunes purchases to Apple-authorized products. Or it can prevent you from using your Kindle's read-aloud function to listen to a book—even if you are blind.[1] It can stop your DVR from recording your favorite show if the copyright holder objects.[2] Through region coding, DRM can stop you from watching a DVD you bought on vacation in London or Tokyo on your TV at home, or from using printer ink purchased abroad. It can even prevent you from skipping commercials and trailers before watching a movie that you own.

Push the limits of these rules, and DRM will push back. At that point, you will discover that your media and devices serve another master. Most of the time, they obey your instructions. But when your commands conflict with those of copyright holders, your stuff betrays you. Perhaps it simply refuses to execute a command, or it may politely inform you that you've exceeded your authorization. DRM might even disable your access or your device altogether. Much like *Fahrenheit 451*'s firemen and their hounds, if rather less imposing, DRM treats our access to the products we lawfully acquire as contingent and impermanent. DRM creates a world in which our purchases aren't in our control. Even our very possession of them is contingent on rules established by an external authority.

Consider the Apple iTunes DRM. For reasons we will discuss, Apple no longer sells music burdened by DRM. But movies and television shows, not to mention apps, are still subject to DRM. Apple spells out the substantive constraints of its DRM in its Usage Rules, which it "reserves the right to modify ... at any time." Your behavior will be "monitored by Apple for compliance purposes," and Apple can "enforce [its] Usage Rules without notice." Those rules provide in part:

- You shall be authorized to use iTunes Products only for personal, noncommercial use.
- You shall be authorized to use iTunes Products on five iTunes-authorized devices at any time.
- You shall be authorized to burn an audio playlist up to seven times.
- You shall not be entitled to burn video iTunes Products or tone iTunes Products.

The specific restrictions imposed by any DRM system are less important than the underlying dynamic they represent. Those restrictions were not created by law. Nothing in the Copyright Act even hints that creating seven audio playlists is lawful, but the eighth crosses the line of infringement.

These rules are not the result of a legislative process or judicial analysis. They enshrine an agreement reached between a retailer and a set of publishers and foisted on the public. Unlike the law, DRM allows for automatic enforcement. We've replaced courts and due process with code and license terms. The law can account for context and tolerate gray areas. It can make exceptions. DRM cannot. It hardwires restrictions on consumer behavior into our devices, robbing them of functionality.

While not nearly as dramatic as flamethrowers and fighting robot dogs, the unilateral right to enforce such restrictions through DRM exerts many of the types of social control that Bradbury feared. Reading, listening, and watching become contingent and surveilled. That system dramatically shifts power and autonomy away from individuals in favor of retailers and rights holders, allowing for enforcement without anything approaching due process.

Imagine if a physical book publisher tried to create similar rules: you can read at night, but not during the day; you can read on the beach, but not on the subway; you can only loan the book to a friend once;[3] and you can't skip the preface.[4] None of us would feel compelled to comply with these demands. No court would call you an infringer, and few would even find an enforceable contract. And the publisher would have no way to find out about our violations or force compliance. But because digital works depend on software and often network connections, copyright owners can construct technologies that impose their whims on us. That power, particularly when it is reinforced through law, creates no shortage of harm.[5]

Smart Cows and Dumb Code

The first efforts to use technology to prevent copying emerged in the early days of the retail computer software industry. In those days, users shared time on mainframes and often wrote their own code. Later, hardware makers viewed software as a tool to drive computer sales. But once software was understood as an independently marketable product, some software makers saw the ease of copying floppy disks as a problem in need of a technological solution. Aside from casual sharing of software among friends and colleagues, swap meets and flea markets began to include computer software—both legitimate and infringing copies among their wares—or warez, if you are of a certain generation. This period of unauthorized distribution had some unexpected consequences; it led to later commercial success for companies that built loyal user bases for future products. It also encouraged innovation. One early video game, *Spacewar!*, was improved and developed in part through unauthorized copying. Even Bill Gates, whose attitude about copying shifted significantly during his tenure at Microsoft, learned to program on unauthorized software.

But understandably, most software companies wanted to cut down on unauthorized copies in order to improve sales. Some attempted to impose speed bumps—relatively minor impediments that would slow down the rate of copying and separate legitimate purchasers who wanted to make backups or share with a few friends from rogue copyists. These early DRM technologies included the linguistically and logistically awkward dongle—a hardware device that had to be inserted into the input/output port of your computer before the software would run. In other cases, DRM was tied to software documentation. For example, on launch a program would prompt the user with a question like, "What is the first word on page 14 of the user manual?" Of course, in response users began exchanging information about how to circumvent these systems, quickly diminishing them to mere annoyances, a pattern that would repeat itself with increasing speed for every DRM system to come.

In the 1990s, this small-scale arms race began to heat up as copying and storing large numbers of software titles became easier because of vast improvements in storage capacity and disk speeds. Coupled with the increasing ease of data transmission over the newly popular Internet and the introduction of peer-to-peer networks like Napster, designed for sharing files with a global community, the perceived need for DRM increased dramatically. Soon copyright holders, who now included Hollywood and the music industry in addition to software makers, invested more and more resources in the hopes of finding a technological fix to the problem of unauthorized copying.

But what proponents of this silver bullet strategy failed to understand, at least initially, is that every DRM system is susceptible to attack. That's theoretically true of every system for obscuring or encrypting information. But DRM, by its very nature, is particularly vulnerable. Normally, if you buy a lock to protect valuables inside your house, you lock the door to outsiders. And you keep the key safe in your pocket, sharing it only with insiders such as family or friends. Outsiders might try to break in, but as long as the lock is well made and they don't get their hands on the key, most will be deterred. With DRM, however, the threat is not from outsiders. It's the insiders rights holders worry will make off with the valuables.

A DRM system that locks out consumers altogether has no value. If Apple's DRM, for example, refused to let you watch a movie after paying for it, even the most fervent Apple loyalists would get their digital movies elsewhere. To let customers watch their movies, Apple has to share the key to this digital lock at some point. Since most of us are not particularly tech savvy, DRM makers share the key, but they hide it somewhere we are unlikely to find it. Cryptography works well in preventing attacks from outsiders who want to intercept a message. It is bound to fail when it is used to protect against misuse by the intended recipient of that message. Given enough time and users, discovering the key that unlocks any given DRM system becomes inevitable and often trivial. There are just too many ways to pick a lock from the inside.

And once a single sophisticated user unlocks a DRM system, it usually doesn't take long until the average person can remove that DRM with the push of a button—or simply download an unencumbered copy from the Internet. The inescapable challenge is what Mike Godwin, a pioneering technology lawyer, once called "the problem of the smart cow."[6] Imagine every single cow in the world locked up in a giant barn with a state-of-the-art lock. No matter how good the lock, eventually one cow will figure out how to escape. Once that cow is out, the other cows—no matter how unskilled at lock picking—are out too. This is the second fundamental flaw in the DRM approach. All it takes is one motivated and skilled person to defeat DRM.

In response, DRM makers have pushed for tighter control over more components of the distribution and playback chains, undermining consumer ownership of their devices and software each step of the way. In the process, DRM has shifted from a largely benign form of authentication to a technologically embodied philosophy that views all users—customers included—as threat vectors, monitoring their actions and enforcing limits on their use of the things they buy. Lawful, mundane consumer behavior—watching movies, listening to music, reading ebooks, or making backup copies—are regulated by code. For many people, the frustration and inconvenience of DRM makes paying for content look like a poor value when DRM-free versions of the same works are widely available. When DRM treats paying customers like criminals, denying them the freedom to use their devices as they see fit, it actually encourages them to infringe.

The Battle for Your Living Room

The transition from a market in which people truly owned and controlled their devices to one tightly regulated by technologies that owe us no real allegiance can be illustrated in the contrast between two familiar technologies: the VCR and the DVD player. The VCR, which hit the U.S. market in the late 1970s, empowered the individual. By owning this device, people could assert a degree of control over their television viewing experience that we take for granted today, but was unheard of at the time. No longer subject to the minor tyranny of broadcast schedules, viewers could record shows and watch them at a time of their choosing and on their own terms.

Faced with this prospect, copyright holders were gripped by a hysteria that seems almost laughable in retrospect, but was earnestly felt at the time. Jack Valenti, president of the Motion Picture Association of America, testified before Congress—with a straight face—that "the VCR is to the American film producer and the American public as the Boston strangler is to the woman home alone."[7] Unable to convince Congress to ban the VCR, a group of movie studios, led by Universal and Disney, sued device maker Sony in 1984.

Universal accused Sony of creating a piracy machine that allowed viewers to make illegal copies of broadcast television programs. Although the VCR was certainly used by some to create infringing stockpiles of shows and movies, it was also commonly used for legitimate purposes like time-shifting—the practice of recording a show to watch later. Some producers, notably PBS mainstay Fred Rogers, had no objections to viewers making recordings, in part because it enhanced consumer control. Rogers understood that once a person bought a VCR, it gave the buyer newfound power. Hollywood had no practical means of controlling what they did with their personal property. And once VCRs were in the market, neither did Sony. It had no way of tracking how people used the device, no knowledge of their choices, and no way to limit them. So the studios urged the U.S. Supreme Court to impose legal control over the design and use of the VCR instead. But the Court refused for two interrelated reasons.

First, the Supreme Court ruled that Sony could not be held responsible for designing and selling the VCR if it could be used for both legitimate and illegitimate purposes. If the design of general-purpose devices like a VCR were controlled by content owners, their functionality would be limited to features that supported Hollywood's business plans at the expense of Sony's interest in producing the most attractive device and the public's interest in controlling its viewing experience. Second, the Supreme Court decided that the VCR was in fact capable of substantial non-infringing uses. It found that many VCR owners used them for time-shifting, which the justices deemed a fair use of broadcast television programming. According to the Court's decision, "the business of supplying the equipment that makes such copying feasible should not be stifled simply because the equipment is used by some individuals to make unauthorized reproductions."

The *Sony* decision, while it benefitted device makers most directly, more subtly vindicated the personal property interests of consumers. It protected their right to acquire general-purpose technology, even if it could be used to infringe. It reaffirmed that we can use the devices we own in non-infringing ways despite the objections of copyright holders. And it protected our living rooms from the kind of surveillance and supervision that would be necessary to police the private use of property.

But the Court's decision in *Sony* sent shock waves through Hollywood. Unlike record labels and publishers, which were accustomed to the unavoidable loss of control that comes from selling copies to the public, the movie industry resisted loosening its grip on its works. Traditionally, studios were able to police public consumption of their works because viewers accessed them through public exhibition, not private taping or sales of copies. They were either shown in movie theaters or over broadcast television and could be tracked accordingly. Even the theaters that showed films didn't own the prints. They typically remained the property of the studio and had to be returned. The idea that an individual could own a copy of *The Good, the Bad and the Ugly*—by recording it from a broadcast, no less—represented a dramatic shift in the power dynamic between copyright holders and consumers.

Even after the VCR was introduced, Hollywood resisted the home video market. Titles were withheld from release or priced extravagantly. In part, that's because Hollywood had its own vision for the home video market, and it didn't include the record button. Universal and Disney supported a competing technology, DiscoVision, which allowed viewers to watch movies at home on large optical discs, predecessors to laser disc and DVD technology. But DiscoVision, unlike the VCR, didn't support recording over-the-air programming or private copying. The design of the technology precluded those lawful behaviors. And as a result, it didn't stand a chance in the market. Instead, Hollywood watched as the VCR emerged as the dominant technology.

But Hollywood learned its lesson. When it came time for home video—by then the movie industry's primary source of revenue—to make the transition to a digital format, the studios threw their weight behind the DVD, in part because it enabled the kind of control over copying that VHS tapes never did. When the DVD was introduced in 1996, virtually every commercial release featured a new DRM system called the Content Scramble System (CSS). By encrypting the contents of DVDs, CSS promised copyright holders much more control over what viewers did with the movies they purchased. Playing a movie requires a secret key, and those secret keys are only available to authorized devices. A device maker who wants to manufacture a DVD player has to get permission from the DVD Copy Control Association, an organization made up of major movie studios, DRM vendors, and Hollywood-friendly DVD manufacturers. Not surprisingly, device makers interested in adding features that Hollywood found threatening—like the ability to make backup copies of DVDs, record televised programming, save small clips for educational use, or even skip previews or advertisements—were not approved.[8] DVDs, like their Blu-ray successors, even use region coding to prevent playback of lawfully purchased and imported discs.

Through their tight control over the design of the DVD format, movie studios achieve the goals that the Supreme Court denied them in *Sony*. If Jack Valenti can get away with analogizing the VCR to the Boston strangler, we feel confident noting the parallels between the DVD player and the Trojan horse. Enticed by the prospect of high-quality digital home video, viewers embraced this new technology. But Hollywood understood the DVD format as a means to infiltrate our living rooms and to turn our home entertainment systems into covert assets. By controlling how we use the devices we assume we own, studios could regulate our private activities, including those outside the scope of any copyright interest.

DRM Goes to Washington

For a time, that strategy worked just as Hollywood had hoped. But like all DRM systems, CSS had a fatal cryptography problem baked into its design. It was only a matter of time until someone found the secret key that unlocked DVDs for licensed and unlicensed players alike. And in 1999, CSS was cracked. We will return to that story shortly, but first we should discuss the steps Hollywood took to prepare for this inevitable outcome. Copyright holders who enthusiastically adopted DRM understood that code alone would never be enough to maintain control over consumer behavior. They needed to enlist the law. But early cases gave copyright holders very little confidence that the courts would reinforce the non-legal rules DRM tried to implement.

Two cases from the period between the *Sony* decision and the introduction of the DVD illustrate the problem. The first involved software company Vault, an early DRM pioneer. It sold a program called Prolok that was intended to stop unauthorized copying of software. To do so, Prolok stored a digital fingerprint on disks and prevented computers from accessing the contents of those disks if the fingerprint was missing. Quaid Software looked at Prolok and saw an opportunity. It created a program called Ramkey and its own storage disks that imitated the Prolok fingerprint and effectively broke Vault's DRM. Quaid sold Ramkey as a backup utility, a function that Prolok prohibited legitimate purchasers of software from performing. Vault sued Quaid, claiming that defeating its DRM violated copyright law.

The Fifth Circuit Court of Appeals, much like the Supreme Court in *Sony*, decided the case by looking to the behavior of users who bought Ramkey. Although some were engaged in infringing distribution, a substantial number used it to make perfectly lawful backup copies of software they owned. Because backing up software you own is legal under the exhaustion principle and specifically section 117 of the Copyright Act, the court was convinced that Quaid couldn't be held responsible for those who used the program for less laudable purposes. Breaking, disabling, or avoiding DRM was not, in itself, illegal.

A few years later, the Ninth Circuit underscored that point in a case brought by Sega, developer of the Genesis video game console. Sega made its own titles for the Genesis system, but also licensed third-party developers to create compatible games. Accolade was unwilling to agree to Sega's licensing terms, which required that Sega, rather than Accolade, manufacture the game cartridges. So instead, Accolade decided to create Genesis games without Sega's approval. It purchased a Genesis console and three Sega game cartridges to discover the interface specifications that allowed the game to communicate with the console. In the process, Accolade discovered Sega's DRM, the trademark security system (TMSS). TMSS was, truth be told, a terrible DRM implementation, even by the low standards of the field. It consisted of little more than a twenty-byte initialization code followed by the letters S–E–G–A. The console searched the game cartridge for the initialization code in a specified location. If it found it, the game would load.[9] If not, gamers saw a blank screen. Accolade copied this lockout code onto its own cartridges to render them compatible with the Genesis hardware.

Sega sued, arguing that Accolade infringed its copyrights by copying its game code in the process of reverse engineering the Genesis interface and implementing TMSS. The court disagreed. Even though Accolade copied Sega's software code in its entirety, it had to in order to figure out how the Genesis communicated with games. That interface information and the TMSS lockout code are beyond the scope of copyright protection, the court held, because they serve a purely functional purpose. Just as in *Vault v. Quaid*, copyright holders were rebuffed in their efforts to use copyright law to stop consumers and competitors from defeating their DRM.

In response to these losses, copyright holders took their case to Congress. In the face of ubiquitous personal computing, new digital media formats, and the popularity of the Internet, they argued that some legislative intervention enshrining the DRM strategy was necessary. Their decade-long effort culminated in two laws—the Audio Home Recording Act (AHRA) and the Digital Millennium Copyright Act (DMCA). The first was narrowly focused on a single technology and is viewed by most legal commenters as a footnote—or perhaps a punchline—in the history of legal regulation of technology. The second was motivated by grander ambitions and has had a lasting impact, though not for the reasons its proponents anticipated.

The AHRA addressed digital audio tape (DAT), which was billed as the next hit format for recorded music after the introduction of CDs. Because DAT allowed for digital copying, copyright holders worried it would lead to widespread infringement. So they convinced Congress—in exchange for granting DAT player manufacturers immunity from copyright infringement claims—to require all DAT players to include DRM.[10] The statute makes it illegal for anyone to manufacture, distribute, or import a DAT player unless it incorporates the Serial Copy Management System (SCMS) or its equivalent. SCMS itself was a simple system that encoded data in DAT recordings that dictated whether additional copies could be made. A record company could, for example, prohibit copies altogether, permit a single copy to be made, or allow copies without restriction. Despite the deep degree of congressional oversight into the design of DAT—or perhaps because of it—the format was a flop in the United States.

Six years later, Congress took up the DRM question again. By the late 1990s, the Internet's potential as a digital marketplace had been recognized but not yet realized. Digital distribution of music and other content was technologically feasible, but because of understandable fears that their products would be freely and widely copied, copyright holders were reluctant to experiment with digital marketplaces. They argued that legal protection for their DRM schemes would give them the confidence necessary to take their first tentative steps toward online marketplaces. Congress responded by passing the Digital Millennium Copyright Act (DMCA) of 1998.

The act had two major components. The first created safe harbors from copyright liability for Internet intermediaries like search engines and ISPs. The second was meant to bolster DRM. Section 1201 of the DMCA made it unlawful to circumvent—that is, bypass, disable, or remove—any technological measure that restricted access to copyrighted material. Essentially, it made breaking DRM illegal, even if doing so did not result in copyright infringement. To return to Mike Godwin's cow parable, it's a rule meant to stop the smart cow. But the DMCA had a strategy for dealing with the dumb cows as well. Section 1201 made it illegal to make or distribute tools or technologies designed to circumvent. So even if some motivated teenager well outside the reach of the U.S. legal system cracked a new DRM scheme, anyone who shared a software program implementing that crack was on the hook as well. These anti-circumvention provisions are subject to a number of narrow and largely ineffective exemptions for activities like reverse engineering and encryption research. The Copyright Office even holds a process every three years to decide on temporary exemptions from these rules. The exemption process forces consumers to bear the heavy burden of establishing that their use of the devices and content they own are lawful. And these exemptions only address potential liability under the DMCA; they offer no protection against traditional copyright infringement claims. Even when exemptions are granted, they are based on existing harms to consumer rights that often highlight the absurd overreach of section 1201. Since their implementation, the anti-circumvention rules have consistently undermined the property relationship between consumers and their stuff by giving legal weight to DRM's intervention.

DRM Goes (Back) to Court

Fresh off their victory in Congress, copyright holders began targeting defendants who made or distributed tools that helped defeat DRM. And their track record in those cases suggests that the DMCA gave copyright holders just what they had asked for.

In the first of these disputes, RealNetworks—an early provider of streaming media content—sued a company called Streambox. RealNetworks developed technology for streaming audio and video files. It relied on a digital "secret handshake" between its server and its player to ensure that third-party applications could not stream RealMedia files. Without the secret handshake, an application was denied access. Streambox developed the "VCR," an application that mimicked the secret handshake to interoperate with RealNetworks' server in order to enable the same sort of time-shifting the Supreme Court okayed in *Sony*. Yet when RealNetworks sued, the court had no trouble finding that the VCR software circumvented RealNetworks' DRM because section 1201 had shifted the balance of power in favor of copyright holders.

After this promising test run, copyright holders set their sights on a bigger target. A Norwegian teenager named Jon Johansen solved the puzzle of CSS, the DRM on DVDs, in 1999. He then wrote a simple program called

DeCSS, which decrypted the content of any DVD. Johansen's goal was to enable DVD playback for users of Linux operating systems. Although there were plenty of licensed DVD player software options for Windows and Mac users, there wasn't a single Linux-compatible program on the market. That meant those Linux users who had lawfully purchased DVDs couldn't watch them on their desktops or laptops.

When DeCSS was subsequently published across the Internet for the world to see, it caught the attention of Eric Corley, a journalist and publisher of *2600: The Hacker Quarterly*. For years, *2600* served as a news outlet and forum for the hacker community, broadly defined. Corley wrote a story about DeCSS and published it on his website, along with the DeCSS code and links to other sites hosting the code. As he would reiterate later in court, Corley added the code to the story because "in a journalistic world, ... you have to show your evidence."[11]

Eight movie studios quickly filed suit against Corley and others, claiming that by publishing DeCSS they trafficked in technologies that circumvented DRM in violation of section 1201. The defendants pointed to a number of non-infringing uses DeCSS made possible. They included the time-shifting so crucial in *Sony* and the backups found lawful in *Vault*. Even more intuitively, they argued that DVD owners had the right to play their discs on their own hardware, just like any other item of personal property. But the court held that the legality of these uses was irrelevant to the question of anti-circumvention liability. That charge did not hinge on any act of infringement, much less the question of substantial non-infringing use. The studios had succeeded in Congress where they had previously failed in the courts. Breaking DRM was unlawful, regardless of the reason. Personal property rights had to give way to copyright owner control. DeCSS was banned, and other courts soon followed suit.[12]

A Failure, at Best

Given these decisive early legal victories, copyright holders could be forgiven for deeming the DMCA a rousing success. But any champagne popping that happened in 2001 would soon prove premature. Even from the perspective of copyright holders, the DMCA's anti-circumvention provisions would be charitably described as a mixed bag. And from the perspective of the public, the DMCA has been an unmitigated disaster. It has jeopardized their privacy and security, impeded innovation and encouraged lock-in, and paved the way for an unprecedented loss of control over the devices they own.

The DMCA has not achieved its stated goals. Almost every major commercial DRM system introduced since it was enacted has been broken, and at an increasingly rapid pace.[13] It took three years from the introduction of DVDs for Johansen to crack CSS—not bad, considering he was twelve at the time CSS was released. By the time Apple launched its iTunes FairPlay DRM in 2003, Johansen had more coding experience under his belt. He circumvented that one within a few months. A few years later, the "unbreakable" BD+ DRM used on Blu-ray discs was broken within a month. Princeton researcher Ed Felten led a team that defeated the music industry's Secure Digital Music Initiative in just a matter of weeks. And game maker Ubisoft's DRM didn't even last a day.[14] It would appear there are just too many smart cows out there. Nor has the DMCA been effective in clamping down on the availability of circumvention tools for the rest of us. Any middle schooler with a smartphone and a few minutes to spare can find them. Or so we've been told. As a result, titles protected by DRM find their way to file sharing networks and other sources of infringing material just as quickly as their non-DRMed counterparts.[15]

The other pitch for the DMCA—that it was necessary to convince copyright holders to risk the waters of digital distribution—turned out to be false. It may have been true in the late 1990s, but it certainly isn't anymore. In fact, DRM often hurts copyright holders as much as it helps them. The market rewards publishers who abandon DRM and punishes those who insist on it.[16] This rise and fall of DRM for digital music downloads is instructive. When Apple launched the iTunes Music Store, the first licensed digital music download store to feature content from the major labels, every track was wrapped in its FairPlay DRM. To hear Steve Jobs tell it, the labels insisted on DRM, and Apple played along. As he wrote in his widely circulated open letter, Thoughts on Music: "When Apple approached these companies to license their music to distribute legally over the Internet, they were extremely cautious and required Apple to protect their music from being illegally copied. The solution was to create a DRM system, which envelopes each song purchased from the iTunes store in special and secret software so that it cannot be played on unauthorized devices."[17] If that's the case, the labels came to regret their insistence once they discovered that DRM benefitted Apple much more than it did copyright holders.

Through a combination of being first, creating a seamless experience for end users, and designing gorgeous devices, Apple soon became the top music retailer in the world. Once it established its dominant position, record labels got their first glimpse of the dangers of DRM. Apple's FairPlay-protected tracks, which music fans collectively spent tens of millions of dollars buying, couldn't be played on competing hardware. The costs of switching from an iPod to a Zune—remember those?—were too high for most people to bear. That hurt competition among device makers and music stores alike. So DRM reinforced Apple's dominant position and weakened the labels' leverage to negotiate over prices, promotions, and other concerns.

Apple was so committed to maintaining this tight control over the retail download market that when one-time DRM crusader RealNetworks created a software tool called Harmony that allowed customers of its competing music store to replicate FairPlay DRM so that tracks purchased from RealNetworks could be loaded on iPods, Apple called them hackers and threatened a DMCA suit. Nearly a decade later, a key Apple engineer even testified that the company's DRM was part of an anti-competitive strategy.[18]

At this point, the labels figured out that the only way out of this mess was to free themselves from the chains of DRM. As Cory Doctorow explains in his book *Information Doesn't Want to Be Free*:

> The labels came to realize that they'd been caught in yet another roach motel: their customers had bought millions of dollars' worth of Apple-locked music, and if the labels left the iTunes Store, the listeners would be hard-pressed to follow them. ... But Amazon offered the labels a lateral move: give up on digital rights management (DRM) software and sell your music as "unprotected" MP3s (which also play on iPods), and you can start to wean your customers off the iTunes Store—or at least weaken its whip-hand over your business. You can set your own pricing, Amazon said; we'll help you with the promos you're looking for, and together we can get some competition into the market. The music industry bought into it, and iTunes dropped DRM not long afterward.[19]

A Disaster, at Worst

As bad as DRM ended up being for the music industry, it has been worse for the public. The lock-in problems that finally convinced the record labels to jettison DRM, for example, were felt acutely by consumers. Many of us are attracted to digital copies because of their relatively low prices. Oddly, digital copies are more than occasionally more expensive than their digital counterparts. But digital usually wins on sticker price. When you can pay $8.99 for an ebook instead of $22.99 for a hardcover, it seems like an easy call. But those low prices are misleading. If you can't resell your books, you can't recoup any of your costs. Tired of the latest dystopian young adult novel? Too bad; you're stuck with it. Before DRM, you could always head down to the local used bookstore or flea market to sell your hardbacks and paperbacks. With DRM, your purchases are tied to a particular technology platform. And that fact raises the cost of switching to a new platform, which in turn means less competition and higher prices.

Because it often requires ongoing communication or authentication, DRM focuses on the present rather than the future. Content is generally most valuable in the initial period after its release.[20] After that, content producers, DRM vendors, and device manufacturers have significantly reduced incentives to respond to concerns about DRM. The short history of DRM is littered with the remains of failed or abandoned protection measures, too often leaving the files they supposedly protected locked away. This leads to problems of unavailability and obsolescence. It also creates serious barriers to preservation. Even without DRM, these are thorny issues. DRM only makes them worse.

Ownership unfettered by DRM encourages innovation, customization, exploration, and repair.[21] This "freedom to tinker," as Ed Felten calls it, allows individuals to contribute to technologies, often in ways that the original manufacturer can't or won't.[22] We see this threat most vividly in connection with the growing class of software-controlled and network-connected devices that make up the so-called Internet of Things. We turn to those in chapter 8. But it's equally true for digital media.

For example, when gamers discovered a way to change the appearance of characters in *Ninja Gaiden, Dead or Alive 3*, and *Dead or Alive Xtreme Beach Volleyball*—admittedly, in some cases to make them appear nude—video game publisher Tecmo sued.[23] These enhancements didn't enable infringement; they could only be used by owners of the games in question. If anything they added to the appeal of and demand for the games. Nonetheless, Tecmo sued the tinkerers who created the modifications and the website that hosted them. The suit was dropped only after the site was taken down. Similarly, Blizzard—the maker of *World of Warcraft*—used the DMCA to target volunteers who developed software that allowed owners of its games to play together online.[24] Years later, Blizzard again relied on the DMCA to stamp out a program called Glider that allowed players to automate repetitive tasks like farming, crafting, and collecting items.[25]

There's no shortage of DRM horror stories, but perhaps the most egregious one took place in 2005, when Sony—once a staunch defender of consumer autonomy—hijacked the computers of nearly two million customers.[26] Sony, gripped by the fear of peer-to-peer infringement, decided it was necessary to prevent CD owners from copying their music to their computers. Writing software that prevents CD ripping is an easy enough task. But since people own their computers, they can decide what software to install and what software to delete. That posed a problem for Sony since, of course, no one would actually want to install a program that crippled their computer and made their CDs less valuable. So Sony needed a way to hide its DRM on customers' computers to prevent them from deleting it.

Sony used a tool called a rootkit, rarely employed by legitimate software developers, to achieve this subterfuge. Rootkits are programs that covertly modify a computer's operating system to blind it to certain files and processes. Once a computer has been compromised by a rootkit, it hides any files that meet certain criteria from both the computer's user and the machine's operating system. So if you open a folder containing a malicious program on a rootkit-infected computer, you won't see it. Or if you use an activity monitor to view the processes currently running on your machine, the hidden program—in this case Sony's DRM—won't be visible.[27]

If all Sony's rootkit did was allow it to hide its copy protection software, that would be bad enough. It's an underhanded move that denies people the right to control what code is running on hardware that they own. But the impact of the rootkit went well beyond DRM. It created security vulnerabilities that left users open to an array of potential attacks. Sony's rootkit was programmed to hide any file or process that began with the prefix "sys." If an attacker wanted to install malicious code on your machine and make sure it went unnoticed by you, your operating system, and your anti-virus software, all they would have to do is add that prefix to the file name.

The range of attacks that could exploit this vulnerability is limitless. The user's data could be altered, deleted, or even held for ransom; the machine could be rendered inoperable; a program could sniff sensitive passwords or collect financial records and other personal data. The list goes on; just use your imagination. The threat was more than theoretical. Within days of the public learning about the rootkit, malicious code leveraging it was spreading across the Internet. A program called Backdoor.Ryknos was transmitted via spam email. Once on a user's system, it opened a communications channel that let the attacker remotely control the user's system—downloading, deleting, and executing files, and gathering and sending information from the compromised machine. So while Sony customers nominally owned their computers, they no longer controlled them.

After independent discovery of the rootkit by at least three different groups of researchers—one of which, in full disclosure, was represented by one of this book's authors—Sony was forced to confront its decision when Mark Russinovich went public with his findings. The response from Sony—then the world's second-largest record label—was underwhelming. First, it downplayed the importance of the rootkit. Thomas Hesse, president of

Global Digital Business, asked, "Most people, I think, don't even know what a rootkit is, so why should they care about it?" Eventually, Sony released tools to uninstall its DRM and the associated rootkit, but those tools caused security concerns of their own. Finally, Sony recalled millions of unsold infected CDs.

The Sony rootkit incident reveals, in an admittedly dramatic fashion, the underlying problems with DRM. The devices and content that consumers reasonably believe they own are guided by secret loyalties and hidden agendas that run counter to consumers' best interests. The things we buy are technologically tethered to their creators, subject to ongoing surveillance, recall, or even destruction. They are not under our control. The rootkit incident also exemplifies the attitude behind DRM. We are not to be trusted, not even with our own computers. And our interests in autonomy, security, and privacy are secondary to rights holders' perceived need for greater control over our behavior. Copyright holders, in condemning infringement, often implore the public to show greater respect for property rights. They might try taking their own advice.

The Effort to Copyright Garage Door Openers

In light of the power the DMCA created to control how people use technology, it was only a matter of time until DRM spread beyond traditional entertainment industries to companies in other sectors. Soon we began to see DRM inside common everyday devices like garage door openers, printers, and coffeemakers. These efforts sought to control consumer behavior not out of a fear of infringement, but as a strategy to reduce competition from firms that wanted to lure customers away with cheaper alternatives. With section 1201 as a powerful new tool, electronics companies had strong incentives to put DRM everywhere.

As DRM-restricted products hit the market, competitors of course found ways to circumvent those controls. Predictably, litigation soon followed. One of the first examples was in 2002, when Chamberlain, a maker of garage door openers, sued its competitor Skylink for making an inexpensive universal remote that could be programmed to open almost any garage door, including those made by Chamberlain. Skylink marketed its remotes as replacements for customers who lost their original Chamberlain remote or as an additional remote for drivers who had second or third vehicles. Chamberlain sold its own replacement remotes for a hefty sum, a market it wanted to control exclusively. So Chamberlain embedded DRM in its garage door opener that required remotes to send a proprietary code before they could open the door. After some experimentation, Skylink discovered the algorithm for this secret code and built it into its own remote.

Chamberlain sued, arguing this was an act of circumvention. The court, recognizing the obvious tension between the personal property rights of owners of garage door openers and the claimed IP rights of Chamberlain, rejected this attempt to expand the reach of the DMCA. As the district court explained, "A homeowner who purchases a Chamberlain [garage door opener] owns it and has a right to use it."[28] The owner of the device can use it in ways that conflict with the prerogative of its manufacturer. On appeal, the Federal Circuit held that claims under section 1201 needed to establish some relationship between the circumvention of DRM and a plausible act of copyright infringement. But that nexus was missing here because "consumers who purchase a product containing a copy of embedded software have the inherent legal right to use that copy of the software."[29] Again, ownership undermined the effort to control how consumers used their devices.

A similar case was filed that same year by printer manufacturer Lexmark against Static Control Components, Inc. (SCC), an aftermarket supplier of replacement parts and ink cartridges. Much like razor companies that make most of their money on replacement blades, Lexmark relied heavily on sales of expensive ink cartridges. SCC competed by selling its own compatible cartridges. Like Chamberlain, Lexmark embedded DRM in its printers and cartridges that prevented its printers from accepting non-Lexmark cartridges. SCC reverse engineered the system and designed its own cartridges to be compatible by fooling Lexmark's DRM. Lexmark sued SCC, claiming that by plugging a rival cartridge into the printer, owners of Lexmark printers were circumventing the company's DRM. But like Chamberlain, Lexmark was rebuffed. As the court explained it, "Purchase of a Lexmark printer ... allows 'access' to the program" that runs the device.[30] So Lexmark's effort to assert ongoing control over that piece of personal property failed.

These cases show that courts are still sensitive to the concerns of private property owners, at least in some circumstances. But they also illustrate the deep desire among device makers to retain control over consumer devices after they have been sold. As we explore in more detail in chapter 8, there are tools aside from the DMCA that they can use to make that vision a reality.

8 The Internet of Things You Don't Own

> The door refused to open. It said, "Five cents, please."
>
> He searched his pockets. No more coins; nothing. "I'll pay you tomorrow," he told the door. Again he tried the knob. Again it remained locked tight. "What I pay you," he informed it, "is in the nature of a gratuity; I don't have to pay you."
>
> "I think otherwise," the door said. "Look in the purchase contract you signed when you bought this conapt."
>
> In his desk drawer he found the contract; since signing it he had found it necessary to refer to the document many times. Sure enough; payment to his door for opening and shutting constituted a mandatory fee. Not a tip.
>
> "You discover I'm right," the door said. It sounded smug.
>
> From the drawer beside the sink Joe Chip got a stainless steel knife; with it he began systematically to unscrew the bolt assembly of his apt's money-gulping door.
>
> "I'll sue you," the door said as the first screw fell out.
>
> Joe Chip said, "I've never been sued by a door. But I guess I can live through it."
>
> —*Ubik* by Philip K. Dick (1969)

Cars, refrigerators, televisions, Barbie dolls. When people buy these everyday objects, they rarely give much thought to whether or not they own them. We pay for them, so we think of them as our property. And historically, with the exception of the occasional lease or rental, we owned our personal possessions. They were ours to use as we saw fit. They were free to be shared, resold, modified, or repaired. That expectation is a deeply held one. When manufacturers tried to leverage the DMCA to control how we used our printers and garage door openers, a big reason courts pushed back was that the effort was so unexpected, so out of step with our understanding of our relationship to the things we buy.

But in the decade or so that followed those first bumbling attempts, we've witnessed a subtler and more effective strategy for convincing people to cede control over everyday purchases. It relies less—or at least less obviously—on DRM and the threat of DMCA liability, and more on the appeal of new product features, and in particular those found in the smart devices that make up the so-called Internet of Things (IoT). IoT has become something of a buzzword, intended to cover a range of devices from smartphones and networked thermostats to self-driving cars and wearable technology. These products generally combine embedded software, network connectivity, microscopic sensors, and large-scale data analytics. In essence, they are computers. As Chief Justice John Roberts recently wrote about mobile phones: "The term 'cell phone' is itself misleading shorthand; many of these devices are in fact minicomputers that also happen to have the capacity to be used as a telephone. They could just as easily be called cameras, video players, rolodexes, calendars, tape recorders, libraries, diaries, albums, televisions, maps, or newspapers. ... It is no exaggeration to say that many of the more than 90 percent of American adults who own a cell phone keep on their person a digital record of nearly every aspect of their lives—from the mundane to the intimate."[1]

That's certainly true of our phones, but it's equally true of so many of the objects of modern life. Your car is a computer with wheels; a plane is a computer with wings; your watch, your child's toys, even your pacemaker are all computers at their core.[2] And as computers, they are susceptible to the same sort of external limitations and controls we've witnessed with previous generations of digital goods. Even if we resist it, we're accustomed to software telling us whether we can watch a digital movie. But what happens when computer code dictates when your light bulbs have to be replaced?[3] Or how fast you can drive?[4] Or whether you can fly your drone in a particular neighborhood?[5] Or what brand of cat litter you can use?[6] What are the social consequences of a smart mattress that collects and analyzes heart rate and breathing data, monitors your movements, and provides you a nightly summary?[7] That's what Samsung's new Sleepsense device promises. Samsung even suggests you track your loved ones by "simply put[ting] the sensor under their mattress ... to receive an analysis of the quality of their sleep via email." What could possibly go wrong?

With so many networked devices in their homes, consumers are relying on home automation hubs—devices that allow them to control their home security systems, lights, garage door openers, and entertainment systems from any place with an Internet connection. The maker of one such device, Revolv, was acquired by Google-owned IoT company Nest in 2014. The Revolv hub sold for $300 and touted a "lifetime subscription" for updates and new features.[8] But in April of 2016, Nest announced it would no longer support the Revolv. What's more, Nest planned to exercise its software-enabled remote control over the devices to render them entirely inoperable. After a May 15 software update, it explained, "The Revolv app won't open and the hub won't work."[9] Alphabet, Google's recently-created parent company, which has its sights set on the self-driving car and medical device markets, decided it was within its rights to reduce a device that consumers bought to nothing more than an overpriced paperweight. Consider that before you buy a Google car.

In this chapter, we look at a small sampling of IoT devices across a wide range of sectors and consider their consequences for ownership and consumer welfare more broadly. In many cases, these technologies offer real benefits. Yet the core cultural and legal shifts they represent strike another blow against ownership in the digital economy.

Jailbreaking Is Not a Crime

The exact origin of the Internet of Things is difficult to pinpoint, but one significant moment in its early history was the introduction of the iPhone on January 9, 2007. Steve Jobs told the assembled crowd, "Today, Apple is going to reinvent the phone."[10] He proceeded to wow them with "a revolutionary mobile phone, a widescreen iPod with touch controls, and a breakthrough Internet communications device" combined in a single product.[11] But like nearly every Apple product, the user experience was carefully choreographed and tightly controlled. iPhone users could only run Apple's iOS. They could only configure the settings Apple allowed them to access. They could only use Apple-approved mobile carriers. And they could only run the applications Apple provided. And later, once Apple launched its App Store, they could only install software that Apple approved—on the basis of opaque and inconsistent standards. What you could do with this remarkably powerful pocket computer depended entirely on what Apple let you do.

This walled-garden approach was a dramatic departure from the approach of general-purpose computers, including Macs, which allowed third-party applications and considerable freedom for user modification. In some ways, Apple's approach to the iPhone was more in line with an earlier phone maker, AT&T. During its decades-long reign as a telecommunications monopolist, AT&T—née Bell Telephone—used a number of strategies to maintain strict control over telephones. As the holder of Alexander Graham Bell's patents, AT&T had total control over the design, production, and distribution of phones. And even after those patents expired, it extended that control by leasing phones rather than selling them, making certain that users didn't acquire property rights in their devices. They also used contractual provisions and legal threats to stamp out innovation, no matter how innocuous.

In the 1940s, AT&T exercised this power by targeting the Hushaphone, a small non-electronic accessory that attached over a telephone receiver to increase privacy and cut down on noise. AT&T forbade its use, and it took nearly a decade of legal battles before the D.C. Circuit rejected that restriction as an "unwarranted interference with the telephone subscriber's right reasonably to use his telephone in ways which are privately beneficial without becoming publicly detrimental."[12] This case, along with the FCC's subsequent *Carterphone* decision, which permitted the attachment of wireless technology to AT&T's phones, paved the way for competition and individual ownership of landline phones.

In some ways, Apple's control over the iPhone is a throwback to these bad old days. But it's one that many consumers happily accepted in exchange for the convenience of integrating all of their online activities into a single device. But not everyone was willing to go along quietly. Apple's restrictions sparked a movement to "jailbreak" iPhones in order to regain some semblance of ownership. "Jailbreaking" refers to the act of eliminating software restrictions and DRM that limit how phone owners can use their devices. With a jailbroken iPhone, you can install any software you choose, replace Apple's operating system with one you prefer, and customize the look and feel of your phone. Jailbreaking is related to, but distinct from, unlocking a mobile phone—the process of removing software locks that prevent you from switching wireless carriers—from AT&T to T-Mobile, for example.

Jailbreaking is not a new practice. Similar homebrew communities formed around other devices long before the iPhone launched, from Xbox hacks to do-it-yourself DVRs.[13] But nothing galvanized that community more than the thought of turning Apple's powerful and ubiquitous product into an open platform. The first iPhone jailbreak was announced on July 10, 2007, just eleven days after the device launched. With each inevitable Apple software update, the jailbreaking community would free that new version within weeks, if not days.[14]

Although it didn't file suit, Apple insisted that jailbreaking was illegal. In 2009, the Electronic Frontier Foundation (EFF) filed a petition with the U.S. Copyright Office requesting formal permission for iPhone owners to jailbreak their devices without fearing anti-circumvention liability. This provoked Apple to explain precisely why jailbreaking should be banned. Despite referring to consumers as "iPhone owners" throughout its filing, Apple asserted that "iPhone users are licensees, not owners, of the copies of iPhone operating software."[15] In other words, when you buy an iPhone, all you own is the physical hardware. The software stored on it that make it work and account for much of its value—from the operating system that enables basic functionality to the built-in Weather and Stocks apps—still belong to Apple.

While perhaps shocking to those with an iPhone in their pocket, this stance was a logical conclusion for Apple, a company with one foot in the software industry and a commitment to controlling the user experience that bordered on zealotry. And because Apple has consistently proven its nearly unrivaled skill as a designer of end user experiences, it succeeded in selling us DRM in the guise of a smart device. It made us believe that a bug was a feature. Consumers recoiled at the idea of these sorts of restrictions when Chamberlain and Lexmark tried to sneak them into our garage door openers and laser printers, but when Jobs offered us the same vision, we lined up to give Apple our money.

Eventually, the Copyright Office ruled in favor of the right to jailbreak phones. However, in doing so, it sidestepped the contentious issue of ownership and focused on jailbreaking as a fair use of Apple's copyrighted iOS. And in 2014, an otherwise hopelessly gridlocked Congress passed, and President Obama signed, the Unlocking Consumer Choice and Wireless Competition Act in response to a petition signed by over 100,000 Americans.[16] Although each of these measures suggests both that people still care deeply about owning their devices and that government can be responsive to those concerns, they are temporary fixes. Both the Copyright Office exemptions and the unlocking legislation expire after three years.

Apple's battle for ownership of our phones signaled the beginning of a much broader shift. Every day, we learn of yet another object that will come with embedded software, location detection sensors, and network connections that limit consumer control and surreptitiously communicate back to its corporate mother ship. And while companies like Apple are slowly making their devices more open and user-configurable as a result of public pressure and competitive threats from open-source mobile operating systems such as Android, whole other areas of our lives are becoming constrained and preconfigured for us, often without our knowledge.

Old MacDonald Licensed a Farm

Farmers have enough to worry about. Banks are coming to foreclose on their land. Locusts are eating their crops. Immigration policy is complicating their hiring practices. And corporate agri-business long ago redefined the economics of their way of life. On top of all of this, today's farmers have to contend with intellectual property.

It began with seeds. For years, Monsanto successfully sold Roundup, an herbicide that helped farmers control weeds and other unwanted vegetation. But Roundup also often damaged the crops themselves, so Monsanto began manufacturing crops resistant to Roundup. It patented so-called Roundup Ready soybeans and later added alfalfa, canola, corn, cotton, and sugar beets to the list of Roundup-resistant products.[17] Initially welcomed by many farmers, some were troubled by Monsanto's claim that its seeds were licensed for a single season, not sold. This meant that no matter how many seeds you saved, they couldn't be replanted the following year, a centuries-old farming practice. Instead, you had to buy new seeds from Monsanto or else contend with pests and less-effective pesticides.[18]

Seed patents were just the beginning of the IP frustrations facing farmers. Software has also found its way onto the farm. The iconic John Deere tractor now contains no less than eight control units—hardware and software components that regulate various functions, ranging from running the engine to adjusting the armrest to operating the hitch.[19] When tractors were purely mechanical, farmers could easily maintain, repair, and modify their own equipment as needed. But now, software stands in their way. That barrier is no accident. Tired of losing revenue to industrious farmers who repaired their own tractors or bargain hunters who took their equipment to an independent repair shop, John Deere decided to force their customers to have their equipment serviced by authorized John Deere dealers. By interposing a software layer between farmers and their tractors, John Deere created a practical hurdle. And by wrapping its software controls in DRM, it created a legal one. A quick glance at the John Deere owner's manual gives you a good indication of the result. Almost any problem—from high coolant temperature to a parking break that's not working or a seat that's too firm—ends the same way, with a trip to the John Deere dealer.[20]

Fed up with John Deere's tactics, a group of farmers petitioned the Copyright Office in February of 2015 for a temporary DMCA exemption, like the one granted to smartphone jailbreakers, that would give them clear legal authority to repair, upgrade, and modify their tractors. John Deere responded with adamant opposition, insisting that tractor owners had no right to look under the digital hood, even if the fix was quick and technically simple.[21] Its argument hinged on ownership. John Deere claimed it owns the software, and not just as an abstract matter of copyright law. It owns the copies of its code embedded in the tractors it sells to farmers, code that is essential to the functioning of the equipment. Farmers, in John Deere's words, merely had "an implied license for the life of the vehicle to operate the vehicle."[22] That means you get to keep driving the tractor you bought from John Deere for tens of thousands of dollars unless and until it tells you otherwise.

John Deere's attitude toward ownership has a number of important implications that typify the core risks presented by the Internet of Things. Most obviously, by denying farmers the right to repair—a right entrenched enough that even patent protection can't disturb it[23]—John Deere has effectively raised the price of its products for farmers. It has also done serious harm to the market for repair services, which are less competitive since farmers have no real choice of mechanics.

Less obvious is the harm locking down tractors can have on innovation. Sir Isaac Newton once said, "If I have seen further, it is by standing on the shoulders of Giants."[24] Of course, Newton borrowed the phrase from Bernard of Chartres, but that only underscores the point. Innovation is, in nearly every instance, an incremental affair. Small contributions add up, sometimes in unexpected ways. As Eric von Hippel describes in *Democratizing Innovation,* user innovation—the process by which users take, modify, and improve upon manufactured goods—is a valuable source of new inventive contributions.[25] Farmers have a long history of just such ingenuity and creative problem solving. As MIT Media Lab researcher Ethan Zuckerman wrote: "If you've flown over the western U.S., you've seen green circles dotting the landscape. This is a model of irrigation that's far more efficient than setting up huge systems of piping—instead, fields are centered on a well and a rolling pipe rotates through the field. This technique was innovated by farmers, but is now in wide use. Ask the companies who manufacture these systems who invented them and they'll claim creation. Show them a photo of the systems developed by farmers, and they'll say, 'But you should have seen their welds—they sucked.'"[26]

Kyle Wiens, a self-described right to repair activist, sees farmers' innovations as part of a broader movement: "In the tech industry, we tend to talk about the exploding Maker Movement as if tinkering is something new. In fact, it's as old as dirt: farmers have been making, building, rebuilding, hacking, and tinkering with their equipment since chickens were feral. I've seen farmers do with rusty harvesters and old welders what modern Makers do with Raspberry Pis and breadboards. There's even a crowdsourced magazine, *Farm Show*, that's catalogued thousands of clever farming inventions over the past three decades."[27]

But if farmers don't own the tools, equipment, and seeds they use on a daily basis, this potentially fertile ground for innovation will lie fallow.

Less Fast, More Furious

John Deere is not alone. Other vehicle manufacturers including Ferrari, Ford, General Motors, and Mercedes-Benz are finding new ways to use technology and law to weaken the property interests of drivers. These efforts take a number of forms—DRM that prevents repair and customization, software that monitors and controls your driving, even restrictions on vehicle resale. The car, once a symbol of freedom and independence, is increasingly a tool for control.

Modern cars, much like John Deere's tractors, rely on dozens of electronic control units. Access to the software code on those control units is necessary for many common repairs. The code is also crucial if a driver wants to change the default tuning of their vehicle to get more horsepower or better fuel efficiency from the engine, for example. Researchers investigating potential safety and security flaws likewise need to look under the metaphorical and literal hood. But control unit code is often inaccessible because carmakers use DRM to keep it under lock and key. Until recently, car owners who broke these software locks risked liability under the DMCA, not to mention voiding their warranties. That meant only people with permission from carmakers can do repairs or research without fear of liability. When car owners asked for permission to access the software that controls their vehicles, GM told the Copyright Office that car purchasers mistakenly "conflate ownership of a vehicle with ownership of the underlying computer software in a vehicle."[28] But when code is inseparable from the essential functions of the vehicle, ownership of a collection of spare parts provides little comfort. Even those who miss their car payments have begun to fear "The Remote Repo Man"—an embedded program that disables automobile operation when the purchaser fails to make their monthly payment with no ability to override, even in emergency circumstances such as rushing to the hospital.[29]

Mercedes-Benz has also followed suit. It offers mbrace, a feature that provides remote vehicle controls, service diagnostics, directions, and vehicle tracking. It also connects to Verizon to provide roadside assistance, emergency help, and even geographic, speed, and temporal restrictions on teenage drivers.[30] These bells and whistles sometimes offer real benefits. But what Mercedes-Benz doesn't advertise is that the code running these features doesn't belong to you. As part of the Terms of Service, Mercedes-Benz insists that "you do not acquire any rights in such software, including any right to use or modify the software[.]"[31] They go on to state: "We may update the software contained in your Vehicle's systems or the Equipment from time to time. We may do this remotely without notifying you first. ... These software updates may affect or erase data that you have previously stored on the Equipment in your Vehicle (for example, specific route or destination information). We assume no responsibility for lost or erased (or otherwise affected) data." In other words, Mercedes can remotely enter your car without notice or consent to update or erase any digital information or feature at any time without taking any responsibility for damage that it might cause. If you own a copy of *1984*, don't store it in your E-Class.

Consumer advocates have pushed back against these efforts, passing a Right to Repair law in Massachusetts and pressuring manufacturers to negotiate a Memorandum of Understanding with aftermarket repair shops and part suppliers that allows those businesses access to diagnostic information for repair and replacement purposes, but not for automobile owners.[32] According to the Automaker's Alliance, "The real issue of concern here is that the sophisticated computers in vehicles are so intertwined that they shouldn't (for security and safety and environmental reasons) be allowed to be tinkered with."[33] But carmakers themselves have a troubled history with their own sophisticated systems. Recent experience suggests that extra eyes reviewing this code might be helpful. Half a million Fords were recalled because of software glitches that prevented their engines from shutting off.[34] And Chrysler recently recalled 1.4 million vehicles because their onboard infotainment systems were vulnerable to hackers. Independent researchers uncovered that flaw when they wirelessly hacked a Jeep driven by a colleague, giving them control over the vehicle's steering and braking.[35]

The notion that car owners can only control the parts of their vehicles that don't yet incorporate software or electronic sensors has serious implications for ownership. Under GM's theory, you can check the air pressure in your tires—for now—but you can't run a diagnostic test on your GPS to make sure you won't end up in a lake. Every time you put gas in your car, there are security, safety, and environmental risks. But unless you live in Oregon or New Jersey, you'd be shocked if you were locked out of your own gas tank. There are pressing public safety concerns associated with operating a car, but they should be addressed by accountable public agencies such as the Department of Transportation, the DMV, state and highway patrols, and the EPA, not through private IP enforcement. As we learned from Volkswagen's "Defeat Device," which allowed it to cheat on federal emissions tests for its diesel vehicles,[36] intellectual property laws that protect embedded software from independent testing and examination have potentially massive consequences for the environment, public confidence, and vehicle resale value.[37]

Fortunately, the Copyright Office, along with the National Telecommunications and Information Administration, agreed to allow vehicle owners to break DRM and access the software in their cars for purposes of security research, personal modification, and repair under a special DMCA exemption.[38] However, as noted, these exemptions last only three years. They are hardly a permanent response to these problems.

Not satisfied with controlling who repairs your vehicle, carmakers want to decide how you drive as well. Ford sells a car equipped with its Intelligent Speed Limiter, which uses onboard cameras to scan road signs for speed limits and then adjusts the fuel to prevent drivers from exceeding posted limits.[39] This sort of technology gives Ford an immense amount of information about your driving habits. As a Ford executive boasted in 2014, "We know everyone who breaks the law, we know when you're doing it. We have GPS in your car, so we know what you're doing."[40] And while there may be good reason to embrace tools that reduce speeding and increase safety, they fundamentally change what it means to own the car you drive and the autonomy we are used to having inside it. As vehicles move toward computer-assisted and self-driving, the incidents of ownership will likely grow more and more distant.

For Ferrari, the antipathy toward owners reached new levels. The company now requires customers to sign a Right of First Refusal Agreement that bars the new owner of a $200,000 car from selling it without checking with Ferrari first.[41] The relevant part of the agreement reads: "Customer hereby grants to Dealer, as a material consideration for the opportunity to purchase [the vehicle], an option to repurchase the [vehicle] at its market value (but in no event more than the original Manufacturer's Suggested Retail Price) at any time within two (2) years of the date of delivery."

Granted, this is a purely contractual restriction. But it provides a window into the sort of control carmakers want over owners. Even more than contracts, software and the DMCA might give it to them.

Free as in Coffee

Those in the free software movement are fond of distinguishing between two ways in which we use the word "free." "Free as in beer" refers to price. "Free as in speech" refers to liberty, the freedom you have to use a thing as you choose.[42] Until recently, you could be confident that if you overheard someone talking about free coffee, it meant Starbucks was running a promotion. But thanks to Keurig, the maker of the popular K-Cup brewing system, conversations about coffee now have to account for questions of liberty as well.

The Keurig saga began in 2012, when several of the coffee company's key patents expired. Those patents covered its pod-based brewing system. Users placed single-serving portions of coffee or other brewed beverages in the machine, hit a button, and got a consistent drink each time. Without patent protection, Keurig had to contend with competition. As it turned out, Keurig wasn't a fan. Rival companies started producing compatible pods and undercutting Keurig's prices. In response, Keurig released new machines featuring "Keurig 2.0 Brewing Technology which reads each lid to deliver on the promise of excellent quality beverages."[43] Marketing speak aside, what that meant was that Keurig's machines would only accept pods embedded with a code that verified your coffee came from a licensed supplier. And it also killed off its generic pod that let you supply your own coffee grounds. If you tried to brew rogue coffee, your Keurig machine greeted you with this cheerful message:

The public reaction was swift and vicious. Angry Facebook posts and irate Amazon reviews flooded the Internet. As Brian Barrett wrote, "A coffee maker limiting your choice of grind seems as out of place as a frying pan dictating your eggs."[44] It didn't take long for competitors to capitalize on this outrage by cracking the Keurig DRM.[45] Coffee drinkers even figured out how to defeat it with a single piece of tape.[46] Soon Keurig was persuaded to reverse course, at least in part. It appears to be sticking to its guns when it comes to blocking pods from competitors, but it announced plans to reintroduce the My K-Cup product that allowed coffee drinkers to fill their own pods. Nonetheless, the company and its investors have paid a price for its overreach. Keurig stock dropped by 10 percent in the wake of the DRM controversy.[47]

The Keurig example shows that people still care deeply about owning and controlling their devices and that they have the potential to make their voices heard in the marketplace. But it also cautions that market pressure is often only partly effective in protecting consumer interests.

Open the Pod Bay Doors, Barbie

At this point, it should come as no surprise that the Internet of Things threatens our sense of control over the devices we purchase. However, those threats aren't limited to intellectual property and DRM; they also include battles for control over information about our behavior and our inner lives. One troubling example is the Wi-Fi-enabled Hello Barbie doll from Mattel. This IoT Barbie looks like many of her predecessors but offers a unique feature. She can engage in conversation with a child and learn about them in the process. Barbie does this by recording her conversations and transmitting them via network connections to ToyTalk, a third-party cloud-based speech recognition service. ToyTalk then uses software and data analytics to analyze those conversations and deliver personalized responses. It's an impressive trick, but the implications for our sense of ownership are quite shocking. For many children, talking to toy dolls is a way to share their unfiltered thoughts, dreams, and fears in a safe, private environment. But according to the terms of the Hello Barbie EULA, ToyTalk and its unnamed partners have wide latitude to make use of information about your child's conversations in ways that few parents would anticipate:

> All information, materials and content ... is owned by ToyTalk or is used with permission. ...
>
> You agree that ToyTalk and its licensors and contractors may use, transcribe and store. ... Recordings and any speech data contained therein, including your

> voice and likeness as may be captured therein, to provide and maintain the ToyTalk App, to develop, tune, test, enhance or improve speech recognition technology and artificial intelligence algorithms, to develop acoustic and language models and for other research and development purposes. ...
>
> By using any Service, you consent to ToyTalk's collection, use and/or disclosure of your personal information as described in this Policy. By allowing other people to use the Service via your account, you are confirming that you have the right to consent on their behalf to ToyTalk's collection, use and disclosure of their personal information as described below.

In other words, ToyTalk claims to own anything you, your child, or even their friends say to Barbie. Conversations with the doll are corporate property. The safety and privacy of a child's bedroom is compromised by the collection, sharing, and commercial use of those conversations. And while these services may offer benefits, they come with significant new risks. Shortly after the IoT-enabled Barbie shipped, security vulnerabilities that could allow hackers to intercept a child's conversations with the doll were revealed.[48] And those worries aren't just hypothetical. Around the same time, VTech—maker of the children's smartwatch Kidizoom and InnoTab mobile device—disclosed that more than six million children had their personal information, including photos and chat messages, stolen from VTech's servers.[49]

Hello Barbie is just the latest example of this trend of networked appliances. Samsung shipped a SmartTV with a default listening mode—and accompanying privacy policy—intended to continuously eavesdrop on viewers and send audio back via the cloud for analysis.[50] In a pitch to investors, Vizio recently touted the fact that its smart televisions will be able to detect any content that users watch, regardless of the source, and use that information to customize advertising and programming.[51] The June smart oven features cameras and software that can recognize the food you cook.[52] Google's Nest thermostat takes a similar approach to learning about you. Amazon's Echo, Apple's Siri, Microsoft's Cortana, and Google Now go a step further by encouraging us to interact with disembodied soothing, friendly, and—by default—female voices. Science tells us that we engage more readily with technology that mimics human interaction. A recent study showed that gamblers risk more on slot machines with humanlike features.[53] Of course, such services have the potential to offer real benefits. But such a service relationship comes not only with divided loyalties but also diminished autonomy. It is very different from owning an object completely and suggests we should be mindful of exactly who controls our relationship with any object we purchase. A person's home may be their castle, but their appliances may belong to someone else.

Our Bodies, Our Servers

As if our connection to the Internet of Things wasn't intimate enough, network-enabled and software-dependent devices are now inside our bodies. When open source advocate Karen Sandler found out at age thirty-one that she could die suddenly from a heart condition, she did what most of us would do. She went to the doctor to fix it. In her case, that meant implanting a pacemaker-defibrillator in her chest to give her heart a jolt in the event it gave out. The device—about the size of an avocado—was literally a life-saving invention. But because it ran proprietary software, Sandler had no way to tell how it worked or how likely it was to fail. As she explained in an interview, "A statistic came out recently that 25 percent of all medical device recalls in the last few years have been due to software failure. When you read these statistics it becomes very personal."[54]

It turns out that Sandler's questions about her pacemaker weren't so easy to answer. Much like Apple and its iPhone, pacemaker manufacturers won't let patients look inside or test the devices they purchase. Nor are you allowed to read the data from your own device while you are at home or on the road—even in the midst of a medical emergency.[55] Instead, you can only access your health data from manufacturer-approved sources. And until recently, you couldn't even test your device to make sure it is functioning correctly or was running the latest software or security update. The reason for such restrictions? According to a filing with the Copyright Office, the Advanced Medical Technology Association "believe[s] that patients have an inherent right to access their own medical data, however, this in and of itself does not necessitate bypass of any intellectual property protections."[56] In other words, even if you own the physical parts of the pacemaker, the manufacturer's copyright trumps any claim you might have to see how it works or what data it collects on you—even when it is implanted inside your body.

Dana Lewis proved what patients can do when they own their devices and control their care. Lewis is a diabetic living in Seattle who relies on a glucose monitor and a handheld wireless device to alert her when her blood sugar is too high or low.[57] Yet Lewis often wasn't able to hear the alarm, especially when she was sleeping. So she and her partner, Scott Leibrand, built a new program that displayed blood sugar levels with new louder alarms and a snooze button. They even added the ability to send the information to other mobile devices, such as Leibrand's Pebble watch. Next they turned to Lewis's insulin regime. Traditionally diabetics control their insulin levels manually. But Lewis and Leibrand began experimenting with the data to devise an algorithm specific to Lewis's needs—something that would automate and adapt based on the data her device was sending out. It could predict her insulin needs thirty, sixty, and even ninety minutes in the future. Eventually they hope to produce an artificial pancreas that will essentially automate this process. No IP law, and certainly not one designed to stop infringers from sharing movies online, should stand in the way of patients adapting equipment they own to keep them alive.

These concerns are not limited to those of us with life-threatening conditions. More than 20 percent of Americans currently use "wearables"—computing devices attached directly to your body.[58] When you buy a Fitbit wearable tracker, its Terms of Sale specifically state that "to the extent the Products contain or consist of software in any form ... such Software is licensed to you, not sold[.] Terms such as 'sell' and 'purchase,' as used in these Terms, apply only to the extent the Products consist of items other than Software."[59] Again all you own is the shell and the components. Everything digital—including physical storage media—belongs to Fitbit. While Fitbit's privacy policy does promise to remove personally identifiable information whenever it shares your records with third parties, it reserves the right to keep everything else indefinitely, even after you delete your account.[60] Every move you make, every step you take, Fitbit will be tracking you. And as Kate Crawford wrote, because the type of information collected by these devices is so personal, and so intimate, it is almost as if the device itself becomes a more authoritative source about us than we are.[61]

Network security has also become an issue for medical devices. From insulin pumps to cochlear implants and powered prosthetic joints, more and more medical devices rely on transmitting medical data to providers through Wi-Fi and Bluetooth protocols.[62] These connections have already opened the door to numerous security issues.[63] Even former Vice President Dick Cheney claims to have switched off the wireless functionality on his own pacemaker to prevent terrorists from hacking it.[64] Fortunately, much like with vehicle security testing, the Copyright Office granted an exemption for testing exterior medical devices and passively testing those that are implanted in ways that don't affect functionality.[65] The ability to innovate and improve these devices, however, remains highly contested.

Karen Sandler's dream of an open source pacemaker may inspire us, but it also presents complications. Open source could allow patients to examine, test, and improve devices in ways far more flexible and permissive than the current proprietary model, but they don't give us autonomy in quite the same way as analog ownership. Instead they offer a future with different, more user-friendly restrictions to navigate. Focusing on medical devices, the argument for individual ownership and control resonates more viscerally. For the rest of the stuff we buy, the stakes may be lower, but the arguments are the same. If you don't own your devices, you can't repair or customize them. You can't innovate with them. And in the end, the products you buy may end up using you more than you use them.

9 Patents and the Ordinary Pursuits of Life

Imagine walking into an auto dealership to buy a new car. You look around, find a promising model, and approach the salesperson to find out the price. They explain: "Oh, have I got a great deal for you! This beauty is 10 percent off the normal sticker price, but there is a small catch. The discount requires that you only use our special 'single-use' tires. That means you can drive on them as much as you like, but once they get a flat or are low on air, you can't repair or refill them yourself; instead you have to order new 'single-use' tires from our online store. Otherwise, I'm afraid you'll infringe our patents."

To some, the notion that the tires you buy couldn't be refilled seems not only inefficient but also absurd. And the idea that refilling them could somehow put you on the hook for patent infringement is even more bizarre. Yet, the most influential patent court in America recently held that these sorts of restrictions on any patented product are legal and enforceable.

The case, *Lexmark v. Impression Products*, involves printer manufacturer Lexmark, a company that has fought for over a decade to stop its customers from buying competing ink cartridges or refilling authorized ones, using the combination of intellectual property rights and DRM. Much like Keurig's coffee scheme, Lexmark's strategy involves selling customers a device, but conditioning their use on the purchase of authorized accessories. You buy the coffee maker, but can only use Keurig's coffee pods; you buy the printer, but can only use Lexmark ink cartridges, and only once.

When Lexmark customers buy one of its ink cartridges, they may discover text on the outside of the packaging telling them that if they open the box, they "agree to return the empty cartridge only to Lexmark for recycling." If they don't accept those terms, they are instructed to "return the unopened package to your point of purchase." If they prefer to pay more, "a regular price cartridge without these terms is available."

Crucially, Lexmark maintains that consumers who violate these restrictions are patent infringers. How could you infringe a patent by reusing a product you paid for? Lexmark argues that even if consumers own their printers, they don't actually unconditionally own the cartridges that contain the ink. Lexmark customers, the company says, merely license the ink cartridges for a single-use, and the act of refilling them infringes its patented technology. If Lexmark is right, a customer's ability to use the printer they own is contingent on Lexmark's permission. A printer is useless without ink, after all. It also means that Lexmark—or any device maker—can leverage its patents over one product to control aftermarkets for related ones.

For Lexmark, attempts to tie its products and accessories together are nothing new. In 2002, as we discussed in chapter 7, the company tried to use the DMCA's anti-circumvention rules to accomplish the same goal. It claimed that the software that exchanged data between its printer and its cartridges contained DRM that protected the printer software from being accessed without authorization. Fortunately, this theory was rejected by the Sixth Circuit Court of Appeals. Recognizing the anti-competitive and anti-consumer impact that such claims could have, one judge wrote: "If we were to adopt Lexmark's reading of the statute, manufacturers could potentially create monopolies for replacement parts simply by using similar, but more creative, lock-out codes. Automobile manufacturers, for example, could control the entire market of replacement parts for their vehicles by including lock-out chips."[1]

That result is clearly not what Congress intended. Despite this rebuke, Lexmark shifted tactics from copyright to patent law, hoping that it could achieve the same control over downstream use via a different legal doctrine. Whether or not it can will depend on how the Supreme Court views the interaction between patent owners and purchasers of patented devices and whether it will recognize the crucial role of patent exhaustion.

Patent Law's Flexible Approach to Exhaustion

Much like copyright exhaustion, patent exhaustion historically has played a central role in curbing attempts to monopolize post-sale uses of patented goods. And while copyright law has focused largely on the statutory embodiment of the first sale rule over the last century, patent law embraces a more flexible common law exhaustion regime. As early as 1852, the Supreme Court held in *Bloomer v. McQuewan* that once a patented good was sold, the patent owner could not interfere with the rights of purchasers to use it "in the ordinary pursuits of life."[2] It announced this rule for many of the reasons we've already discussed, emphasizing personal property: "[W]hen the machine passes to the hands of the purchaser, it is no longer within the limits of the [patent] monopoly. ... The implement or machine becomes [the purchaser's] private, individual property."[3]

But what does it mean to use a digital good in the ordinary pursuits of life? At the very least, you would expect to be able to use a device for its intended purpose, to repair it, and to transfer it. But according to Lexmark and many manufacturers of the things that make up the Internet of Things, those customary uses depend on patent holder permission. Yet the history of patent exhaustion directly contradicts that assertion. A mere twenty-six years after the *Bloomer* decision, the Supreme Court stated with unusual clarity that the purchase of a patented product "carrie[s] with it the right to the use of that machine so long as it [is] capable of use."[4] Equally clear was that this right to use one's purchases trumped any attempt by patent owners to impose restrictions on how or where one could use the product. The case, *Adams v. Burke*, involved a coffin maker who sold special coffin lids to undertakers in the Boston area. The lid-maker tried to impose a territorial limit that barred use of its lids outside of a ten-mile radius. But the Supreme Court rejected this attempt under the rule of patent exhaustion. According to the Court, it is part of "the essential nature of things" that when a patent holder sells a device, it "parts with the right to restrict [its] use."[5]

Two years after *Adams v. Burke*, the Supreme Court again struck down an attempt by a patent holder to impose post-sale restrictions on a purchaser. In *Keeler v. Standard Folding-Bed Co.* the Court said in no uncertain terms that "one who buys patented articles of manufacture from one authorized to sell them becomes possessed of an absolute property in such articles, unrestricted in time or place." The Court was convinced that "the inconvenience and annoyance to the public that an opposite conclusion would occasion are too obvious to require illustration."[6]

One of the most controversial technological innovations of all time—the electric chair—helps illustrate the power exhaustion gives product owners to ignore the objections of the patent holder. On August 6, 1890, the Westinghouse Electric Company's Alternating Current Dynamo was put to a novel use—killing a human being. One year earlier, William Kemmler, a twenty-eight-year-old seller of fruits and vegetables in Buffalo, N.Y., had wandered home drunkenly and murdered his lover Tillie with an ax. After the attack, Kemmler allegedly told a bystander, "I've done it and I expect to take the rope."[7] But technology was one step ahead of Kemmler. Earlier that year, the State of New York became the first to authorize electric execution, and on May 13, 1889, Kemmler was sentenced to "die by electricity."[8]

During the previous decade, the use of alternating current, or AC, to kill humans had been the center of a contentious political and publicity war between two of the most famous Americans of the Gilded Age—industrialist millionaire George Westinghouse and Thomas Edison, inventor of AC's rival, direct current (DC). Despite Edison's ingenuity, Westinghouse, with the brief help of the electrical savant Nikola Tesla, had been gaining ground quickly in the electric power market. Desperate to promote DC, Edison sought to highlight the dangers of AC. The electric chair thus became the perfect symbol of what the Edison camp was calling "the killing current."[9]

But Westinghouse owned patents on the key components of AC and refused to license them for use in execution. Still, Edison's agents would not be deterred. Recognizing that Westinghouse's patent rights could be exhausted for individual dynamos after their initial sale, they reached out to one of Westinghouse's disgruntled licensees and secretly bought up used Westinghouse alternators in order to build the first three electric chairs in history. Westinghouse objected, but patent exhaustion prevented him from controlling the devices after they entered the stream of commerce. In this way, patent exhaustion helped Edison demonstrate how deadly AC could be using Westinghouse's own patented products.[10] Had Westinghouse been able to restrict the use of his dynamos like Lexmark and other patent holders do today, the debate about the electric chair would have focused on hypotheticals instead of the gruesome reality that became the center of the discussion.

It's easy to see why patent holders might object to these sorts of unauthorized uses. But the Supreme Court understood that patent exhaustion furthered important interests in terms of both competition policy and consumer protection. So it resisted attempted end runs around patent exhaustion by companies that used techniques like those adopted today by Lexmark. In two 1917 cases, the Court rejected efforts to attach strings to the sale of patented devices. In the first, *Straus v. Victor Talking Mach. Co.*, the Court struck down a licensing notice that attempted to both fix the resale price of phonograph players and also force purchasers to use the patent holder's records and needles exclusively. The Court held that this was nothing more than an attempt "to sell property for a full price, and yet to place restraints upon its further alienation." Such tactics, the Court explained, are "hateful to the law" and "obnoxious to the public interest."

The second case, *Motion Picture Patents Co. v. Universal Film Manufacturing Co.*, involved Edison's patented film projectors. Edison was happy to use Westinghouse dynamos over the patent holder's objections, but when it came to his own patents, Edison had a very different view. His projectors were extremely popular, but Edison quickly realized that the real money was in selling film reels, which he had also patented. So he attached a large steel plate to each of his projectors that asserted that they could only be used with Edison reels. After his patent on the reel expired, the defendants decided to make their own compatible film reels for use with Edison's projector. Edison sued them and their customers, claiming that use of the new reels with the patented projector violated the restriction stamped on the side of the device.

For Edison—much like Lexmark and Keurig—tethering a device that customers are likely to buy only once to a consumable accessory product like ink, coffee, or reels of film looked like a savvy business model. By locking out competitors, Edison could keep the more lucrative film reel market to himself. However, the Supreme Court's *Motion Picture Patents Co.* decision ultimately rejected Edison's attempt to trump patent exhaustion. It explained that "the primary purpose of our patent laws is not the creation of private fortunes for the owners of patents, but is 'to promote the progress of science and the useful arts.'" As a result, "the right to vend is exhausted by a single, unconditional sale, the article sold being thereby carried outside the monopoly of the patent law and rendered free of every restriction which the vendor may attempt to put upon it."[11]

This "single-recovery" approach is rooted in the basic purposes of IP policy. By limiting patent holders to a single profit per sale, it maximizes the incentives to distribute new inventions to as many people as possible and at the same time encourages purchasers to fully utilize the products they buy. It also avoids idiosyncratic arrangements of rights that impose high information costs on purchasers.[12] Limiting patent holders to a single recovery also guards against the abuse that would likely occur if patent holders were granted ongoing control over products released into the stream of commerce: "A restriction which would give to the plaintiff such a potential power for evil ... is plainly void because wholly without the scope and purpose of our patent laws, and because, if sustained, it would be gravely injurious to that public interest, which we have seen is more a favorite of the law than is the promotion of private fortunes."[13]

Just as the Supreme Court recently reiterated in the copyright context, the foundation of patent exhaustion is "the common law's refusal to permit restraints on the alienation of chattels."[14] In *Kirtsaeng v. John Wiley & Sons, Inc.*—the used textbook case—the Court emphasized "the importance of leaving buyers of goods free to compete with each other when reselling or otherwise disposing of these goods." In large part, exhaustion is a reflection of the fact that "American law ... has generally thought that competition, including freedom to resell, can work to the advantage of the consumer."[15] That's just as true for patented devices as it is for copyrighted works.

The Return of Edison's Label

In light of such powerful statements in favor of exhaustion over the last 150 years, you might wonder how companies like Lexmark can continue to insist that they have the right to control how we can use the products we buy. If the Supreme Court rejected Edison's film projector label, what makes the conditions on Lexmark's printer cartridges any different? The answer comes in the form of a case decided by the Court of Appeals for the Federal Circuit—the court that handles all patent appeals in the United States—just a couple of decades ago. In *Mallinckrodt v. Medipart*,[16] the Federal Circuit tried to rewrite the history of patent exhaustion. And it may have succeeded.

But to understand how *Mallinckrodt* departed from the settled law of patent exhaustion, we need to step back fifty years. Before the age of semiconductors, electronics relied on vacuum tubes to control electric current. *General Talking Pictures v. Western Electronic*[17] involved patents on vacuum tube amplifiers. Western Electric, a subsidiary of AT&T, licensed its patents to the Transformer Company to manufacture tubes for private home use. But Western Electric reserved the right to license their vacuum tube amplifier patents for commercial use—in movie theaters, for instance. In other words, the Transformer Company was authorized by the patent holder to make and sell devices only for the private home market; it wasn't allowed to make or sell devices for the commercial market. The Transformer Company sold those devices to General Talking Pictures, which supplied them to movie theaters, drawing Western Electric's ire and eventually a lawsuit.

On its face, the facts of the case seemed straightforward. The manufacturer knowingly made and sold the invention without a valid license. To manufacture a patent device, you need permission from the patent holder. Otherwise you are an infringer. Western Electric could have granted the Transformer Company a license that permitted it to make as many devices as it could, for whatever purpose, and sell them to whomever it chose. But that's not what Western Electric did. It granted a license "expressly confined to the right to manufacture and sell the patented amplifiers for radio amateur reception, radio experimental reception, and home broadcast reception." Exhaustion requires a sale authorized by the patent holder. But when the Transformer Company sold devices to General Talking Pictures, those sales were anything but authorized. They were expressly forbidden, as both parties knew. As the Supreme Court put it, "The patent owner did not sell to [General Talking Pictures] the amplifiers in question or authorize the Transformer Company to sell them or any amplifiers for use in theaters or any other commercial use."[18] So those unauthorized sales were infringing. For more than fifty years, this seemed like the obvious and accepted holding of the case.

In 1992, however, the Federal Circuit decided *Mallinckrodt*. There, the court adopted a novel interpretation of *General Talking Pictures*. Rather than being a case about unauthorized sales, the Federal Circuit interpreted *General Talking Pictures* as a case about "conditional" sales. But that reading was inconsistent with more than a century of Supreme Court law denying patent holders the right to place conditions on patented objects that had been legitimately sold.

Mallinckrodt involved patented aerosol delivery devices used in hospitals to dispense a mist for use in diagnostic lung X-rays. The devices cost about $10 to make but were sold for closer to $50. The patent holder, Mallinckrodt, labeled the devices "Single use only," allegedly to encourage their proper disposal as "biohazardous waste."[19] Of course, Mallinckrodt was surely aware that the single use restriction would boost device sales as well. But rather than buying new devices after every use, some hospitals sent the depleted devices to defendant Medipart, who recharged them and sent them back to the hospitals for reuse. Mallinckrodt sued, claiming that its "Single use only" label trumped exhaustion and precluded hospitals from working with Medipart to recondition the devices.

The judges hearing the case faced the question that all such cases pose—how to balance the interests of intellectual property owners with the rights of consumers and aftermarket competitors. Perhaps influenced by the adoption of licenses in the software industry, the Federal Circuit sided with the patent owner, finding that unless the restrictions placed on purchasers somehow rose to the level of an antitrust violation for monopolizing an industry, consumers and competitors could simply choose not to purchase the restricted goods.

When confronted with the Supreme Court precedents, including *Motion Picture Patents*, the judges effectively shrugged their shoulders. They equated those precedents with antiquated notions of consumer protection and outdated economics. Instead the court favored the "freedom of contract"—the idea that parties should be free to strike whatever bargain they think is best—ignoring the fact that similar restrictions on the sale of patented goods had been rejected for the previous hundred years and failing to distinguish between a breach of contract and the infringement of a patent. Whatever the agreement between Mallinckrodt and the hospitals, Medipart—the defendant in the case—never agreed to anything.

With this ideology in hand, the Federal Circuit embraced a distorted rewriting of *General Talking Pictures*. That case, it suggested, stands for the proposition that a "valid condition of the sale" of patented goods bars exhaustion and limits what the purchaser can do with a product. However, the court failed to explain adequately how Mallinckrodt's "Single use only" label differed from Edison's or the restrictions in *Bloomer v. McQuewan, Adams v. Burke, Keeler v. Standard Folding-Bed Co.*, or *Straus v. Victor Talking Machine Co.* Its reading ignored the longstanding hostility toward such restrictions. It also leads to the conclusion that, simply by including a label on a product, a patent holder can eliminate exhaustion and the rights that go along with it.

In the aftermath of *Mallinckrodt v. Medipart*, the world of patent exhaustion has been in disarray for over two decades. What exactly is the difference between a "valid condition of the sale" and an unenforceable post-sale restriction? Can you refill your coffee cups and printer cartridges or not?

Even the Supreme Court's most recent forays into patent exhaustion suffer from their own lack of clarity. In a 2008 case, LG Electronics sued Quanta Computer for using LG's patented semiconductor chip technology in Quanta products. Quanta purchased the chips from a licensed retailer, Intel. So by all accounts, it owned the chips outright. However, LG argued that its contract with Intel specified that Intel had the right to sell LG-licensed chips, but Intel's customers couldn't actually *use* those chips without a separate patent license from LG.

Legally, this argument has some appeal. Patent rights can be divided up into bits and pieces, just like real property. So why, LG argued, shouldn't it be allowed to sell the right to make a semiconductor chip to Intel and then sell the rights to use the chip to Intel's customers? Among other things, that strategy might allow for price discrimination. The buyers of the chips might be willing to pay considerably different sums for the privilege of using them.

But the Supreme Court wisely rejected those arguments. It understood that "the initial authorized sale of a patented item terminates all patent rights to that item."[20] LG couldn't maintain control over the use of their patented devices after an authorized sale. When a patented item is lawfully made and sold, "there is no restriction on [its] use to be implied for the benefit of the patentee."[21] After all, what would the point of buying something be if you couldn't use it? Moreover, the Court recognized that if it allowed patentees to avoid exhaustion through the use of artful drafting by their lawyers, they could "shield practically any patented item from exhaustion."

Because the Supreme Court rebuffed LG's attempt to condition downstream use of its products, many have read the Court's opinion in *Quanta* as undermining the foundation of *Mallinckrodt*. In holding that LG's patent rights were exhausted, the *Quanta* decision acknowledged that contract law, not patent law, is the proper framework for enforcing post-sale restrictions. The Court suggested that even though an authorized sale occurred, a breach of contract claim might survive.[22] For the Court, potential remedies under the contract were separate from the question of exhaustion.

Mallinckrodt had assumed the opposite, concluding that "[u]nless the condition violates some other law or policy (in the patent field, notably the misuse or antitrust law), private parties retain the freedom to contract concerning conditions of sale," and thereby retain their patent rights as long as the restriction is "reasonably within the patent grant."[23] But *Quanta* found that if there is an authorized sale of an article, no amount of contracting can change the fact that the patent owner's rights in the article have been exhausted.

Self-Replicating Technologies and the Puzzle of the Perpetual Copying Machine

Similar to copyright, patent exhaustion has also been complicated by technological advances, and in particular technologies where reproduction or replication is simple or even self-executing. In 2013, the Supreme Court again revisited the doctrine of patent exhaustion, this time in relation to genetically modified soybean seeds. Monsanto owned patents on these seeds, and sued farmers who saved seeds from prior seasons and replanted them, claiming this infringed the exclusive right to "make" their patented products. Bowman, one of these farmers, argued that patent rights in the seeds were exhausted when farmers bought the original batch, and any subsequent seeds that came from the harvested plants were subject to exhaustion as well. Seeds, he argued, are naturally "self-replicating"; they grew themselves. The Court rejected Bowman's arguments, including the so-called "blame-the-bean" defense, but it noted that its holding was limited to the facts of the case; other technologies might, in fact, self-replicate "outside of the purchaser's control" or that self-replication might be "a necessary but incidental step in using the item for another purpose." In noting this, the Court cited to section 117 of the Copyright Act, which you'll recall allows for the creation of essential step copies and modifications of software programs.

Another self-replication case will likely arise in the near future, presenting courts with even greater challenges for balancing intellectual and personal property rights. The fact that farmers are confronting them in the context of their seeds and combine harvesters shows exactly how uncertain ownership of technology has become. After all, if we can't easily enjoy the ordinary pursuits of life on the farm, where can we?

Selling Globally, Exhausting Locally

Just as copyright law confronted the question of international exhaustion in *Kirtsaeng*, patent law is trying to decide what to make of sales of patented devices that occur outside of the United States. If foreign sales trigger exhaustion, products bought overseas—sometimes at much lower prices—can be imported into the U.S. market. If they don't, global commerce becomes fragmented and complex for any product containing a patented technology.

The confusion over this issue began in 2001, when despite a long line of cases finding exhaustion could be triggered by any authorized sale in the world, the Federal Circuit found just the opposite in *Jazz Photo Corp. v. International Trade Commission*.[24] The case involved Fuji Photo, a company that patented the disposable camera—a novelty hit at weddings, graduations, and birthday parties before ubiquitous camera phones. Event planners would hand them out to attendees, who would snap photos during the festivities and then leave the cameras behind to be developed as a batch. Fuji's competitors saw an opportunity to take the used cameras, ship them oversees to be restocked with film, and then import them back into the United States for sale. Fuji sued, claiming that despite lawfully purchasing the used cameras, these companies infringed Fuji's patent. Refurbishing them, Fuji argued, was the equivalent of making a new patented product.

The Federal Circuit held that, for cameras purchased in the United States, patent exhaustion applied and refurbishing was perfectly legal. However, for cameras bought abroad, refurbishing was infringement. Why the difference? According to the court, "United States patent rights are not exhausted by products of foreign provenance." In support, the decision cited a single Supreme Court case from 1890, *Boesch v. Graff*, for the proposition that "a lawful foreign purchase does not obviate the need for license from the United States patentee before importation into and sale in the United States."[25]

The *Boesch* case, however, said no such thing. In that case, the patentees held patents on lamp-burners in both Germany and the United States. The defendant purchased lamp-burners in Germany, not from the patent holder, but from Hecht. Hecht was permitted, as a "prior user" under German law, to make and sell the lamp-burners. The question in the case was whether the defendant could resell the lamp-burners purchased from Hecht in the United States. The Supreme Court concluded that sales made by Hecht—as opposed to sales made by the patent holder—did not trigger exhaustion.[26] Again, the fundamental requirement of exhaustion is an authorized transfer of the object by the *rights holder*. So in *Boesch*, there was no exhaustion—not because the sale occurred overseas, but because the defendants didn't purchase the products from the patent holder. Rather than see the case for what it was, a holding that refused to apply exhaustion in the absence of an authorized sale, the Federal Circuit's reading of *Boesch* gave patent holders the worldwide right to geographically discriminate, a major departure from over a hundred years of exhaustion precedent.[27]

The Supreme Court's decision in *Kirtsaeng*, rejecting territorial limits on copyright exhaustion, casts serious doubt on the continued viability of *Jazz Photo*. As we noted earlier, *Kirtsaeng* stressed the common law roots of exhaustion, which made no territorial distinctions. Those roots are shared by patent exhaustion. *Kirtsaeng* warned of the absurd results that a strict national exhaustion regime could inflict on commerce. The Court noted that under such a rule, for example, cars made overseas couldn't be resold by their domestic owners because they contained copyrighted code. That's equally true of the thousands of patented components in your vehicle or smartphone. *Kirtsaeng* also dismissed the notion that copyright holders were entitled to segment markets geographically: "The Constitution's language nowhere suggests that [copyright] should include a right to divide markets or a concomitant right to charge different purchasers different prices for the same book, say to increase or to maximize gain. Neither, to our knowledge, did any Founder make any such suggestion. We have found no precedent suggesting a legal preference for interpretations of copyright statutes that would provide for market divisions."[28]

Despite the cold reception it received at the Supreme Court, that's precisely the argument patent holders make against international exhaustion. If foreign sales trigger exhaustion, it throws a wrench in their carefully laid plans. One response to this argument is to say, as the Supreme Court did in *Kirtsaeng*, "too bad." Patents, like copyrights, do not entitle their holders to control all valuable uses of their products. Those rights have limits, and patent holders have to live with them.

Yet just as John Wiley pointed to the positive impact price discrimination could have on students in developing countries who need cheap textbooks, patent holders have told their own, even more compelling story of the upside of market segmentation. Instead of cheap books, patent holders point to cheap pharmaceuticals. Citizens in developed countries like the United States can generally afford to pay much more for a product than those in poorer or less developed nations. By charging rich countries more, drug companies can charge poor countries less. And often that's what happens. For example, one 2010 study examined the difference in international drug prices and found that in the top five countries, the prices were almost five times as high as they are in the bottom five countries.[29] The result, patent holders claim, is a net increase in access to potentially lifesaving medicine.

Putting aside the fact that the pharmaceutical industry doesn't tell us much about the market for smartphones or ink cartridges, there are reasons to doubt the accuracy of this simple story.[30] No doubt, some patients in developing economies benefit from price discrimination. But not all do. Drug companies are sometimes tempted to take advantage of the vast disparities of wealth within poor countries by selling their products at high prices to a lucrative minority. Many drugs are still unaffordable in developing countries despite strict bans on exporting them.[31] And countries like the United States have their own problems with wealth inequality. When rich countries supply subsidies through high consumer prices, the poor in those countries don't fare well.[32] Ultimately, while pharmaceutical companies feel public pressure to keep drug prices low in the developing world, the goal of price discrimination is not to increase social welfare but to maximize profits. Just like textbook publishers, large drug companies enjoy extraordinary profits. In 2014, Forbes reported Pfizer's profits at an astounding 42 percent.[33] That's not a company that sets prices on the basis of social welfare.

Even accepting the argument that price discrimination should be encouraged, it is far from clear that it depends on restricting patent exhaustion. If Pfizer wants to charge different prices in different countries, it has plenty of tools at its disposal. As a patent holder, it has incredibly strong bargaining power and can insist on contract terms that restrict imports into the United States. Product tracking technologies have improved dramatically, so detecting breaches through customs inspections and other forms of commercial surveillance are much more likely today than even a decade ago. Aside from avoiding breach, pharmaceutical manufacturers and distributors have reputational incentives to keep up their end of such a deal. There's also the nonpatent regulation of pharmaceuticals to consider. For prescription drugs made in the United States and sold abroad, the FDA explicitly prohibits re-importation, even if those drugs aren't patented. In short, it's not clear that a national exhaustion rule is necessary for price discrimination.

Exhaustion's End?

And so we arrive at *Lexmark v. Impression Products*, decided in February 2016.[34] In that case, the majority of a twelve-judge panel of the Federal Circuit upheld Lexmark's right to restrict resale of printer cartridges the company had authorized for sale but had marked with a "Single Use" label, reaffirming the court's commitment to the flawed reasoning of *Mallinckrodt*. And at the same time, it stuck to its guns on the *Jazz Photo* rule, holding that authorized sales outside of the United States do not trigger exhaustion, despite the Supreme Court's holding in *Kirtsaeng*.

In its opinion, the Federal Circuit articulated a radical rewriting of not only patent exhaustion, but the nature of consumer property interests. According to the court, when a patentee sells you a product, the extent of your rights to use and enjoy that product are entirely dependent on the wishes of the patent holder. The sale, in the Federal Circuit's understanding, represents an implied license to use or transfer the product—not unlike John Deere's theory that farmers merely have an implied license to drive their tractors. But if the patent holder announces—through a label, sticker, or some other means—its desire to restrict your behavior with respect to things you've bought, your property rights have been unilaterally redefined. But that's not how property law works. Nor is it how courts have historically understood exhaustion. Property rights in chattels, even patent-protected ones, are not subject to the attachment of these kinds of puppeteer's strings. If consumers' rights to use and transfer the things they buy were really contingent on implied permission, those rights can be taken away at any time. That is not what ownership looks like.

Aside from its mangling of the post-sale restrictions question, the Federal Circuit also sidestepped *Kirtsaeng*'s universal exhaustion rule. Instead of acknowledging the shared common law origins of patent and copyright exhaustion, the court took pains to distinguish the two bodies of law. It pointed out that despite rewards abroad, international exhaustion could also deny patent holders the reward of an initial sale in the United States, but that is equally true for copyright holders under *Kirtsaeng*. It pointed out that the scope and availability of patent protection varies between countries, but copyright laws are not uniform either. It also invoked the principle of territoriality—the notion that the United States cannot regulate conduct outside our borders—but exhaustion does not require extraterritorial application of the law, merely the recognition of facts that occurred abroad. And in any case, *Kirtsaeng* rejected that same argument in the copyright context. Perhaps more effectively, the Federal Circuit relied on the fact that *Kirtsaeng* turned on the Supreme Court's interpretation of section 109 of the Copyright Act and that the Patent Act has no equivalent statutory provision, so the Court's textual analysis is inapplicable. But the Supreme Court's reading of the statute was heavily influenced by the centuries of personal property common law that preceded it, a history that should have informed the Federal Circuit's understanding of patent exhaustion just as it informed the Supreme Court's understanding of copyright exhaustion. As we write, petitions have already been filed in the Supreme Court. If the Court agrees to hear the case, we hope it will recognize some measure of respect for the rights of buyers. Otherwise, ownership in the digital economy will be at even greater risk.

Whether it's international exhaustion, self-replicating technologies, or attempted post-sale restrictions, patent law is confronting the question at the heart of this book—what do you own when you buy a product? It's the question the Supreme Court asked more than 150 years ago in its first patent exhaustion decision and the one the Federal Circuit continues to struggle to adequately answer. In the next and final chapter, we turn to some ways courts and lawmakers might answer the question more clearly and more fairly.

10 Ownership's Uncertain Future

So far, this book has explained the shifting relationship between consumers and the products they acquire, and the factors driving those changes. Overlapping developments in the law, technology, and the marketplace have undermined our sense of ownership of the digital and tangible goods that surround us. In this concluding chapter, we have two objectives. First, we will explain why—despite the many benefits of rentals, subscriptions, and sharing—an economy in which ownership disappears is a cause for concern. And second, we will offer a sketch of the kinds of interventions we think could help safeguard ownership and the many interests it serves. Our goal is not to turn back the clock or forestall innovation. Instead, we want to highlight the consequences of undermining ownership in hopes of preserving meaningful choice and the many benefits of personal property.

Ownership, Sharing, and Choice

A future that deemphasizes ownership is not only inevitable, it's already here. The explosive growth of streaming services like Netflix and Spotify, accompanied by plummeting physical media sales, tells only a small part of the story. Likewise, digital media and devices hobbled by license restrictions and DRM are already facts of life. Although important in their own right, these are all examples of a much broader and deeper cultural shift away from ownership.

We see evidence of this transformation in the emergence of the so-called "sharing economy." For those unfamiliar with the term, it refers broadly to services and business models that enable individuals and organizations to share, rent, and reuse resources, often enabled by technology. If you've ever gotten a ride in an Uber or spent the night in an Airbnb rental, you've taken part in the sharing economy. The range of goods and services in the sharing economy is staggering. In addition to rides and apartments, there are platforms for renting parking spots, bicycles, private planes, and clothes. Other platforms help neighbors share tools and household goods. Leftover-Swap and EatWith even apply the sharing model to meals.

In everyday English, "sharing" implies something given freely. A few sites like NeighborGoods and Streetbank actually facilitate sharing in the literal sense. And that sort of sharing, of course, is premised on individual ownership. You usually can't lend something that you don't own. But many of the services lumped together under the "sharing economy" moniker are premised on short-term, for-profit rentals. Most are built around the exchange of money for temporary access. Some services rely on a distributed network of individual owners connecting to end users through a technology platform. Others depend on a single provider that coordinates the needs of lots of users.

The rapid growth of some of these efforts has attracted lots of attention. But we rented cars, stayed in hotels, and endured rented bowling shoes long before the first iPhone app. So what is it—if anything—that makes the sharing economy "disruptive"? For one, we see nonownership models moving from out-of-the-ordinary circumstances, like renting a car on vacation, to the everyday convenience, like ride sharing on your commute to work.

In large part, the expansion of temporary-access models is a function of technology. Before everyone had a smartphone in their pocket, the transaction costs of renting your bike for a few hours were prohibitive. By making it easier for owners and users to connect, technology enables more efficient use of existing resources. Cars, for example, are parked most of the time. If you can reliably press a button to summon a ride that takes you where you need to go—especially if it's cheaper—why own a car? Of course, public transit users have been asking and answering that question for decades.

The decline of ownership is also a function of reduced wealth, particularly among millennials. Post-recession, young people are less likely than previous generations to prioritize traditional financial milestones like buying a car or house. The number of young people who own cars or homes has dropped significantly in recent years.[1] Increasingly, ownership looks like a luxury they can't afford. In that sense, Uber isn't much different from Spotify. If you have money to spend, you can own a car or a record collection. Or you could spend a lot less for access to services that provide rides and music.

People are understandably attracted to the appeal of a lower sticker price. But sometimes the price tag fails to reflect hidden costs and other risks. Ownership has long-term upsides for individuals and society as a whole that aren't always readily apparent—in terms of privacy, autonomy, and competition, among others. That's not to say we should do away with new models of allocating and sharing resources, or that we should favor incumbents at all costs. But we need to be fully aware of the bargains we are striking.

There are losers in the sharing economy, and they aren't just legacy taxi companies and expensive hotels. The savings Airbnb users realize and the company's profits are in part the result of externalities—costs that Airbnb and its users aren't bearing. In cities big and small, there is evidence that Airbnb contributes to rent increases for residents.[2] As more housing units are devoted to the sharing economy, fewer are available for locals to rent. Long-term renters have even been evicted to make room for vacationers.[3]

The unseen costs of the sharing economy are also borne by the increasing number of workers classified as independent contractors. By insisting on that classification for its drivers, Uber—currently valued at over $50 billion—avoids paying the minimum wage, payroll taxes, health insurance, unemployment benefits, and workers compensation for the vast majority of its workers.[4] That cost shifting isn't apparent to Uber users. All they see is the few dollars they saved compared to a taxi and the free bottle of water.

These problems are hardly insurmountable. They are largely the function of business models that have outpaced the law. But the law will catch up. Airbnb is under increased scrutiny by local authorities. And Uber is in the midst of litigation over the employment status of its drivers. In all likelihood, the savings that flow from the efficiency introduced by the innovations of the sharing economy are here to stay. But the externalities they've relied on so far are not.

Temporary-access models also leave us vulnerable to fluctuations in price. When we depend on resources owned by others, we have to pay the prices they demand. Uber's controversial surge-pricing model illustrates the point. When demand for rides is high—because of a sporting event or a hostage crisis[5]—Uber responds by increasing prices, by as much as eight times the normal fare.[6] Uber defends its policy on the grounds that higher prices should convince more drivers to hit the road to meet consumer demand. And competition puts some limits on price. If you don't want to pay Uber's inflated rates, you can take a rate-regulated taxi, or public transit, or drive your own car. That's true for now, at least. But once we head down the path of temporary access, it might be hard to reverse course. In a world of licensed, robot-driven cars—a world that may soon be upon us—it might not be so easy to drive yourself.

As the benefits of temporary access models—primarily price and convenience—contribute to their spiking popularity, we worry about the long-term impact on choice. If we neglect the physical and legal infrastructure of ownership, we may see it disappear. Manufacturers are building cars, electronics, and other devices that we can't truly own; DRM keeps them loyal to another master. Publishers are launching digital-only imprints.[7] And despite the recent resurgence of vinyl, there are only twenty pressing plants in operation in the United States, and they struggle to meet demand.[8] Without the means of production and delivery, ownership becomes more expensive, if not impossible. Once the record stores are gone and the CD plants have closed, the competitive checks on the price of services like Spotify weaken.

Just as important, when we trade ownership for access, we sacrifice reliability. Most of us have had the experience of realizing the movie we were hoping to watch has disappeared from our Netflix queue.[9] Titles come and go all the time as licenses expire and new deals are negotiated. Or maybe your favorite album is pulled from your subscription music service because the artist signed an exclusive deal with a competitor. These are minor inconveniences, to be sure, but they highlight the contrast between a model in which the consumer has control and one in which control is entrusted to a third party. More troublingly, works can disappear altogether. In a world without individual ownership, a publisher could pull a controversial book, movie, or album from the handful of subscription services, and it would be like it never existed.

And when you don't own your devices, you lose control over the kinds of uses you can make of them. So far, limitations on use have for the most part treated everyone the same: "you cannot lend this ebook," or "your rental period is twenty-four hours." But as technology reduces the costs of monitoring and valuing individual behavior, we are likely to see increasingly fine-grained, individualized use-based restrictions. Imagine your reasonably-hip crossover vehicle alerting you after your third after-school stop, "I'm sorry; you've reached your limit of daily passenger drop-offs. Would you like to upgrade your vehicle plan to CarPoolPro?" As if that weren't indignity enough, your carmaker's pricing algorithm—relying on information it has gathered about property values in your neighborhood, your driving patterns, and your in-car search history—predicts exactly how much you are willing to pay for the privilege of dropping off that last cranky first grader.[10] This is exactly the goal of the price and geographic discrimination tactics we have discussed throughout the book—to divide our lives into individual transactions and charge as much as we are willing to pay for each one. Shifting away from ownership is an essential step toward that future.

Temporary-access models are not inherently harmful. Whether they are the right choice for any particular consumer will depend on a number of factors—the type of resource, the use they want to make of it, how they value that use, their income, their desire for durability, and a constellation of other considerations. Someone who embraces Netflix might treasure their record collection. And a commuter who could never part with their bicycle might enthusiastically sign up for a robot car service. There's no one solution for all occasions. What's important is that we can choose between ownership and temporary access depending on our needs.

Physical formats and business models come and go. Columbia House is actually relaunching for vinyl. What separates the shift we are witnessing today from, say, the demise of the illuminated manuscript is that the conceptual and legal framework of ownership is crumbling. In the rest of this chapter, we look at ways we can preserve the choice to own.

Avenues for Legal Reform

One way to safeguard ownership is through changes in the law. Law is a powerful tool for regulating markets and protecting the interests of individuals. The notion of intellectual property itself is a creation of the law—a legislatively crafted reprieve from competition. Considered in this light, balancing the rights of IP holders and consumers is inescapable, regardless of who comes out ahead. The law can favor rights holders, as it often does. Or we can leverage law as a tool to help preserve the benefits of ownership. But there is no single legal fix for the full range of issues we've detailed in this book.

Of course, legal change faces hurdles. The judicial process is slow, and courts tend to be incremental in their innovations. The legislative process suffers from its own difficulties. In the current political climate, Congress can barely avoid regular government shutdowns, let alone reach agreement over substantive policy changes. And historically, copyright lawmaking has done a poor job of taking the public interest into account.[11] But despite these stumbling blocks, large-scale copyright reform is underway. Spurred by Register of Copyrights Maria Pallante's call for the "next great Copyright Act," Congress has taken some initial steps toward rethinking the law for our digital economy.[12] What will come of that effort remains to be seen.

That said, there are steps lawmakers, courts, and regulators can take if they understand the problems facing consumers and are motivated to address them. Some are a matter of enforcing existing law. Others depend on courts interpreting aspects of the law in ways that are more sensitive to the threats to ownership. Others require legislative change.

Preventing False Promises of Ownership

One partial solution we've already discussed centers on making sure we have accurate information when we choose between the ownership and access models. Today, consumers have no good way to distinguish between a sale that confers meaningful rights of ownership and a license that imposes all manner of restrictions on their use of a product. Both are labeled "sales," and we are encouraged to "buy" and "own" with nary a mention of the special meaning those words are intended to convey in the digital context. And as we've shown, lots of people are in fact deceived. If consumers want to choose rentals or subscriptions, they should be free to do so. But they shouldn't be fooled into sacrificing ownership by misleading language. Since consumers can't easily challenge these practices themselves, courts are unlikely to ever hear a false advertising case challenging the "Buy Now" lie. That leaves responsibility for policing these abuses in the hands of the Federal Trade Commission, an enforcement agency that is more than capable of targeting these behaviors. We urge the FTC to investigate the "Buy Now" button and encourage retailers to adopt a short notice that clearly identifies what buyers can and can't do with digital goods.

Limiting Form Contracts

As we argued in chapter 4, courts should stop analyzing licenses as contracts and regard them instead as pure grants of permission. But even if courts insist on the license-as-contract framework, bringing contract law to its senses is another way to chip away at the edges of the ownership problem. Licenses that attempt to redefine the relationship between consumers and their purchases rely on the legal fiction of freely negotiated agreements. But that fiction does not reflect reality and should be cast aside by the courts. Being put on notice of a contractual term is not the same thing as agreeing to one. And the law's imposition of a "duty to read" that holds people accountable for terms they didn't examine is a holdover from a bygone era when purchases were not routinely accompanied by thousands of words of legal limits. Today, no one can be expected to read the overwhelming onslaught of complex terms and conditions—many of which are longer than this chapter—that consumers confront in the digital environment. It's time courts stop pretending we should.

Ideally, courts should protect individuals by reining in the worst offenses of EULAs through the contract law doctrine of unconscionability. For an agreement to be unconscionable, it has to be shockingly one-sided as a result of unequal bargaining power between the parties. Courts could refuse to enforce licenses that claim to eliminate ownership in a transaction that has all the hallmarks of sale by deeming them unconscionable. Those take-it-or-leave-it agreements are certainly one-sided, and some provisions, like ones that unilaterally reserve the right to terminate or alter the agreement certainly appear unreasonable. But courts aren't eager to find unconscionability out of a reluctance to intervene in the marketplace. But when the marketplace is riddled with unread terms and misleading marketing, intervention is necessary.

Perhaps contract law's embrace of marketplace dynamics and its overreliance on notice rather than requiring meaningful assent means that it is incapable of addressing concerns about EULAs. If so, we need to look elsewhere for solutions. One promising tactic is the FTC's approach to advertising disclosures. Faced with print, TV, and online ads for a range of products that "quoted prices, but didn't adequately disclose the strings that were attached," the FTC announced a policy that requires clear and conspicuous disclosure of the relevant terms.[13] Disclosures are evaluated using the FTC's *Four Ps*: prominence, presentation, placement, and proximity.[14] Prominence requires that the disclosure be easily readable, and the fine print of a EULA doesn't suffice. Presentation looks at whether the disclosure is written in a way that will be easily understood; dense legalese fails that test. Placement considers whether the disclosure appears in a place consumers are likely to look; presumably that disqualifies EULAs that no one reads. And proximity examines the relationship between the disclosure and the claim that it modifies; a "Buy Now" button isn't particularly proximate to a revelation that "this product is licensed not sold" disclosed in a linked EULA. If these sorts of standards were used to scrutinize most EULAs on digital media and devices, those terms would fail miserably.

The EU's Unfair Terms in Consumer Contracts Directive offers another example of a legal regime that requires extra scrutiny for certain types of consumer contracts.[15] In general, when contract terms are not individually negotiated, the Directive considers them unfair to the extent they lead to a significant imbalance in the rights and obligations of consumers and merchants. The Directive also provides a number of specific examples of unfair terms familiar from many EULAs for digital goods. They include terms that: limit the legal rights of the consumer in the event of nonperformance or inadequate performance by the seller; obligate the consumer even where the seller does not perform its obligations or where its performance is optional; allow the seller to terminate a contract without reasonable notice;

and permit the seller to alter the terms of the contract or the characteristics of a product unilaterally.

Under the Directive, unfair terms are not binding. In addition, the Directive requires that terms are drafted in plain language and that ambiguities be interpreted in consumers' favor. EU states are required to enforce these standards under their national laws. If the United States adopted similarly rigorous standards for form contracts, some of the most egregious abuses in EULAs could be avoided.

Another way to rein in contracts would be to strengthen the doctrine of copyright preemption. When a state law—like contract—conflicts or interferes with a federal law—like copyright—federal law wins. Theoretically, courts could rule that EULAs that are inconsistent with copyright law—if they deny owners the right to transfer, for example—are unenforceable. However copyright preemption is rare because courts mostly focus their attention on the issue of whether or not the rights defined by a contract overlap with the rights of a federal copyright. They almost never ask whether a contract interferes with the rights of consumers of the copyrighted work. We think that view misunderstands not only the relationship between licenses and copyright, but also the fundamental purpose of our IP system.

Freeing Owners from DRM

Even without contractual restrictions, the machine-code limitations imposed by DRM remain a major barrier to owners' control over their property. DRM can constrain how owners use digital media and software-embedded devices in ways that were impossible in the predigital era. Looking at the anti-circumvention provisions of the DMCA nearly twenty years after their creation, we see them as a major policy misstep. They have stifled innovation and competition, fragmented markets, impeded research, stymied educators, and compromised security. Unintended consequences aside, those provisions have proven ineffective and unnecessary when it comes to their stated purposes. DRM has rarely prevented or even slowed infringement, and at least one study has shown it actually reduces sales.[16] As the music download market has shown ever since Apple abandoned DRM, technological protections are not needed to convince copyright holders to sell content online or for fans to buy it. And the most outrageous abuses of DRM—in our garage doors, printers, and vehicles—bear no connection to copyright infringement at all.

We see very little downside to scrapping section 1201 altogether.[17] But if policymakers are insistent on keeping it, there are two partial solutions to the problems it creates for ownership. Courts have already held that unless an act of circumvention bears some "reasonable relationship" to an act of copyright infringement, the DMCA cannot stand in the way of breaking digital locks.[18] Where a use is protected under the fair use doctrine or section 117—which articulates exhaustion rights in computer software—some courts have found that "critical nexus" to infringement is missing.[19] It would be a modest expansion of existing law to recognize that circumvention undertaken for purely personal use—like repairing your tractor or reading a book on an unsupported device—or to enable a transfer of ownership—like giving away your digital movie to a friend—should also be immune from DMCA liability.

A cleaner solution, but one that would require legislative action, would exempt owners from section 1201 altogether. If you own a digital good—whether it's a movie or a car—software locks shouldn't stand in the way of you accessing or using your property in ways that are otherwise lawful. DRM that prevents you from reading an ebook on a new device or diagnosing your sluggish engine does not protect any legitimate copyright holder interest. And someone who hacks a car's operating system in order to sell infringing copies to competing carmakers would still be liable for copyright infringement.[20] The legal dragnet of the DMCA, however, ensnares more average users than it does determined hackers. We'd also exempt makers of tools that enable circumvention from liability to the extent those tools are primarily designed for and used by owners of digital goods. We shouldn't expect every ebook reader to figure out on their own how to make an iBook work on a Kindle.

That's not to say the DMCA's anti-circumvention rules would be entirely toothless. Under this proposal, DRM could still enforce the clearly articulated limits of rentals and subscriptions. Since renters and subscribers are not owners, rights holders could prohibit circumventing software code that limits access to and use of a digital good. So if a user exceeds the twenty-four-hour rental period or fails to pay the monthly subscription fee, self-executing code could cut off their access. We think this distinction is an intuitive one. While we might find them annoying, we accept limitations on our use of products we rent. But that's quite distinct from code that controls what we can do with digital content or devices that we own.

Reinvigorating Patent Exhaustion

As this book is being written, the Supreme Court is weighing the outcome of the Federal Circuit's decision in *Lexmark v. Impression Products*, the case that decided it was illegal for you to refill your ink cartridges. There are two distinct questions presented to the Court in that case—first, whether patent holders can restrict how a purchaser uses a product after it has been sold, and second, whether authorized foreign sales should be treated the same as domestic ones in terms of triggering exhaustion. With respect to both of these questions, we urge the Supreme Court to correct the Federal Circuit's efforts in recent decades to rewrite the law of patent exhaustion. Instead, it should return to the well-established rules in patent law that rejected post-sale restrictions and artificial geographic limits on exhaustion and reaffirm the positions it outlined in *Quanta Computer v. LG* and *Kirtsaeng v. John Wiley & Sons*.

The other alternative is legislation. Intervention by Congress is preferable to living with the Federal Circuit's current perspective on exhaustion. But even acknowledging that court's missteps over the past two decades, we remain confident that a flexible common law approach that allows courts to apply the fundamental principles of exhaustion to evolving facts is the best way to resolve disputes in a fast-changing market.

Reforming Copyright Law

Addressing the core of the ownership crisis requires changes to copyright law. Those changes could be achieved through legislation; they could take the form of courts embracing a more expansive view of exhaustion; or they might require both.

A number of legislative solutions have been proposed over the years. One of the earliest came in 1997 when then-Representative Rick Boucher introduced the Digital Era Copyright Enhancement Act.[21] That bill, proposed as an alternative to what became the DMCA, would have amended the statutory first sale rule to permit the owner of a lawfully made copy "in a digital format" to reproduce, perform, display, and distribute the copyrighted work to a single recipient, so long as the owner "erases or destroys" their copy "at substantially the same time." In essence, the bill would have allowed an owner to transfer their interest in a digital good, even if that meant making copies, so long as they didn't keep any for later use. The bill was prescient and boasted fifty bipartisan cosponsors. But it failed after copyright holders worried that it would lead to widespread infringement since, they argued, no existing technology could have ensured compliance by copy owners.[22]

More recently Representative Blake Farenthold introduced the You Own Devices Act or YODA in 2015.[23] That bill addressed the more narrow but pressing problem of transfers of software-enabled devices. As we've seen, many device makers insist that purchasers don't own the software built into their phones, cars, and appliances. That could render resale of those devices an act of infringement. YODA would prevent that absurd result. Under the bill, when a computer program enables a product to operate, the owner of that product "is entitled to transfer an authorized copy of the computer program, or the right to obtain such copy, when the owner sells, leases, or otherwise transfers" the product. That right cannot be waived by contract. So far, these sensible amendments have garnered little support. However, in a promising turn, Senators Grassley and Leahy have recently asked the Copyright Office to study the extent to which copyright law undermines legitimate uses of software-enabled devices by consumers.[24]

If Congress fails to act, the courts should step in. After all, it was the courts, not Congress, which created the exhaustion principle in the first place. And courts tended to be more flexible in its application. Before the first sale doctrine was codified in the Copyright Act, courts interpreted exhaustion more broadly. Exhaustion went beyond the right to distribute a copy. It included the rights to modify it and to make reproductions to repair or restore a copy. We think that same sort of context-sensitive adjudication could allow courts to apply the basic principle of exhaustion to digital goods. Statutory changes tend to be narrow and rigid, but judicial change offers flexibility. But given the text of sections 109 and 117, courts understandably feel constrained when it comes to pushing the accepted boundaries of copyright exhaustion. So although we believe courts have the power today to revive a broader approach to exhaustion, they might need some additional encouragement from Congress.

Ideally, Congress would endorse a less rigid approach to exhaustion that can accommodate the realities of the digital marketplace, but one that would empower courts to engage in the sort of careful balancing of competing interests they are particularly well suited to do. This solution parallels the history of copyright law's other crucial limitation—fair use. That doctrine got its start as judge-created common law. When it was eventually incorporated into the Copyright Act of 1976, Congress wisely adopted a flexible framework of four nonexclusive factors to guide judges in fair use cases.[25] And although fair use has not always been a model of perfect clarity, this framework has allowed copyright law to adapt to new technologies, market conditions, and uses of works with reasonable predictability.[26]

So what would a multifactor framework for exhaustion look like? Of course, the first question a court would need to decide is whether a particular consumer is an owner or not. We outlined the sorts of considerations we think courts should take into account there in chapter 4. They include the length of possession by the consumer, whether payment is one-time or ongoing; and how the transaction is characterized to the public. In short, we think a one-time payment made in exchange for permanent or open-ended possession of or access to a digital good—whether it's a tangible copy or an intangible asset—results in ownership. That's especially true when the transaction is characterized using words like "sale," "buy," or "own." So when you click "Buy Now" and pay $9.99 for a digital movie, you own it, even if no permanent copy is ever stored on your device. And when you exchange cash for a coffee maker, you own both the hardware and the software embedded in it.

Assuming the court is dealing with an owner, next it has to decide whether the actions they have taken fall within the scope of exhaustion. In other words, did they exercise their property rights or did they make a use reserved for the copyright holder. Right now, the prevailing view is that—with the exception of computer software—exhaustion covers distribution and little else.[27] For the reasons we've described, digital exhaustion should permit acts of reproduction, and even the creation of derivative works, to the extent necessary to enable the transfer of rights from one owner to another. In some cases that transfer of rights will involve moving a particular copy from one location to another. In other cases, it will require the creation of new copies. And in others, it will be a matter of associating permission to access some intangible resource with a new user. But no matter the mechanics, exhaustion can't be a pretext for a digital free-for-all, figuratively or literally. Establishing limits on what an owner can do with their purchase is crucial.

In making this determination, courts should consider:

1. The extent to which the owner parted with possession of or access to the digital good;
2. The extent to which the use deprives rights holders of a fair return; and
3. Whether the owner has materially altered the underlying expression of the copyrighted work.

The first factor identifies the central feature of a lawful sale, rental, or gift. When you transfer your rights in a resource, whether permanently or temporarily, you lose access to it. If you sell your car, you don't get to keep driving it. If you rent out your spare bedroom for the week—unless you are the worst Airbnb host ever—you don't get to sleep in it. With tangible goods this result is partly dictated by physics. That's the nature of rivalry. But it's also a function of the legal construct of property. For digital goods no less than physical ones, a transfer of rights can't lead to an increase in the number of people simultaneously enjoying the work. So if an owner doesn't give up their rights—if, for example, they "sell" their digital record collection but listen to a backup copy—their behavior isn't protected by exhaustion.

The extent to which they are required to give up their access depends on what rights they have and the kind of transfer at issue. If you bought two copies of a favorite book, you could keep one and lend one. The same should be true of digital goods. And different types of transfers demand different degrees of loss. A gift or a sale dictates a permanent loss; lending or rental would entail a temporary one. Sometimes, there will be hard questions. Should the owner of a multivolume work—Julia Child's two-volume *Mastering the Art of French Cooking,* for example—be allowed to lend one ebook but keep the other? What does it mean for a digital work to be published in two volumes? Does it depend on whether the two volumes were sold separately or as a bundle? As a practical matter, physical copies make these sorts of questions easier. But the definition of *the work* is a problem copyright law struggles to answer in a number of contexts.

The second factor looks at the impact of the use on the economic incentives of rights holders. Part of the justification for exhaustion is that a sale offers a strong indication that the rights holder has been fairly compensated.[28] Rights holders set the sale price, after all. So complaints about reduced revenue from secondary markets can't be enough to overcome exhaustion. If that were the case, libraries and used record stores would be outlawed. But there are important differences between digital and analog goods that, in some circumstances, could divert enough revenue from rights holders to undermine the incentive structure of copyright. Courts need a way to identify and address those circumstances.

In the analog world, copyright holders could rely on the inherent limits of physical goods to curb the impact of exhaustion. You can only lend a favorite novel to so many friends before wear and tear, and the occasional spilled drink, take their toll. But digital goods are different. They can be transferred far more quickly and at much lower cost. And in the short term, digital goods are more durable. An ebook of that same novel can be read a thousand times with no wear and tear. Over the long term, however, digital goods face their own challenges. Hardware and software evolve quickly. Storage media and file formats that were popular just a decade or two ago can quickly become ancient relics, leaving digital works practically inaccessible long before their analog counterparts.[29] And hard drives, flash memory, and CD-ROMs all degrade over time. Nonetheless, courts need to be sure that resale and lending don't undermine the basic incentive structure of copyright law in light of the characteristics of digital goods.

Imagine an online community for ebook lending. Thousands or even millions of users sync their ebook collections with this service, enabling them to search for books they want to read. When you borrow a book from user A, it gets transferred to your device, and no one else, including user A, can access that book. So far, this doesn't sound much different from analog book borrowing. But imagine that the ebook lending platform knows whether the book is being read at any given moment. If not, it lets another user check out user A's book. With a large enough user base, this far more efficient lending system could guarantee that once sold, an ebook would never have a wasted cycle. Someone, somewhere would be reading it every second of every day—without any fear of broken bindings or torn pages—all thanks to a single purchase by user A. If exhaustion allowed this sort of system, we could see sales and publisher revenue plummet, perhaps below the threshold of profitability. One response would be to dramatically raise prices to offset lost sales, or to abandon sales altogether and move to a subscription model. None of these are outcomes copyright law should encourage.

This hypothetical illustrates why we cannot simply port the exhaustion rules of the analog world over to the digital marketplace. Courts need a way to assess the impact of uses potentially enabled by exhaustion to see if they cause undue harms to rights holders. Our second factor is meant to do exactly that. What counts as undue harm is admittedly a tough question. That's largely because policymakers—despite the rhetoric of incentives—have consistently failed to measure the economic impact of IP rights, much less begin a serious conversation about the ideal level of incentives. The more incentive, the better, they seem to believe. But no serious assessment could support that conclusion. Limitations on exclusive rights—and by extension, incentives—like exhaustion are essential. But how far should they go?

There are two ways to identify uses that give rise to undue harm to copyright holder interests. First, we could leave it to the courts. That's what the fair use doctrine has done. The fourth fair use factor requires courts to consider "the effect of the use upon the potential market for or value of the copyrighted work."[30] If a use causes enough harm, that fact weighs heavily against fair use. The statute doesn't tell courts exactly how much or precisely what sort of harm, but over decades of common law reasoning, courts have developed a reasonably clear understanding of market harm. There's no reason they couldn't do the same when it comes to exhaustion.

If that creates too much uncertainty, either Congress or the courts could identify categories of use that are beyond the scope of digital exhaustion.

The Copyright Act already adopts this approach. It prohibits rental, lease, or lending of music and software—though not video games. Noncommercial lending remains lawful, however.[31] And section 117 sets up a fairly detailed set of rules regarding archival, adaptation, and necessary-step copies, and their transfer. Policymakers could do something similar with digital goods. They could permit resale and gifts, but prohibit rental and lending. Or they could distinguish between private and public lending, allowing you to lend an ebook to a friend, but not a stranger. They could even try to replicate some of the transaction costs of the analog world by limiting the number of times an owner could lend their digital content, or how frequently. Given these options, there is no reason to insist that exhaustion can't be reconciled with digital distribution.

The third factor is simply meant to prevent exhaustion from becoming a back door to changes to a work that are better considered under fair use. Sometimes an owner will want to modify their copy for compatibility purposes—to make their iBooks work on a Kindle, for example. Exhaustion should permit that, just as section 117 allows owners to adapt their computer programs to work with new hardware and software. But exhaustion isn't the right framework to analyze the legality of remixing or making other expressive changes to a work.

Our approach allows courts to directly and transparently assess the impact of exhaustion on owner and rights holder interests. It will cement a set of ownership rights that are rooted in longstanding property rules, but are attuned to the differences of the digital economy. Those rights will be dictated by law, not privately drafted fiat. We think this framework addresses the concerns at the heart of the exhaustion debate and will prove adaptable to the inevitable changes in the marketplace.

But there's another objection copyright holders raise to exhaustion that we should address. They worry that exhaustion could be used as a cover for widespread acts of infringement. Armed with the right to lend and resell digital goods—the worry goes—purchasers will share copies with friends and strangers, or even resell multiple copies of a single purchase. The concern isn't that exhaustion would permit or excuse that behavior, but that it would somehow makes it easier to get away with. Given the ease of copying online, they say, there's too much infringement already. Any change that would increase the risk of infringing behavior is a nonstarter.

We understand this hesitation, but don't find it particularly persuasive. First, those who want to acquire copyrighted material without paying for it already have ample opportunity. Second, exhaustion is what's called an affirmative defense. That means the doctrine identifies a set of behaviors—lending or reselling, for example—that would normally be unlawful, but are excused for one reason or another. And as an affirmative defense, exhaustion places the burden of proof on defendants. So if a copyright holder suspects that some person or entity is going beyond the lawful scope of exhaustion, it is that person or entity who has to prove that their actions are legal. They'd have to show that the initial purchase was lawful, that they were owners at the time of the resale or lending, and that their actions were of the sort permitted by the exhaustion rule. If they can't prove that, they are infringers.

Granted, monitoring secondary markets and finding potential infringers imposes costs on copyright holders that they would rather avoid. But the possibility of infringement around the edges of resale markets is hardly new. Nothing stopped you from burning a backup copy of your CD collection before selling it to the used record store, or taping your LPs to reel-to-reel in an earlier era. And the expectation that copyright holders bear the costs of rooting out infringers is well established both offline and on. Surprise visits to used record stores and flea markets to find unauthorized copies were just a cost of doing business in the analog world. And courts have reaffirmed that obligation again and again on the Internet—whether it's Tiffany's duty to locate counterfeit jewelry on eBay[32] or the burden of copyright holders to identify unauthorized videos on YouTube.[33]

Despite the dire predictions and fears of copyright holders, consumers outside of the United States are already allowed to resell their digital goods. And the sky has not yet fallen. In a case called *UsedSoft GmbH v. Oracle*, the European Court of Justice ruled that purchasers are entitled to resell the software they buy, even when it is delivered digitally and subject to a restrictive license agreement.[34] Oracle sued UsedSoft for allowing users to purchase second-hand software. Oracle claimed that when UsedSoft users downloaded the software, they illegally reproduced the code. The court disagreed, explaining that since the software was originally purchased lawfully from Oracle, exhaustion applied. The court understood that exchanging a one-time payment for the right to download and use the software was a sale. But that sale was not tied to any particular download or copy, rather it was tied to the right to use the software.

More recently, a Dutch court extended the UsedSoft rationale to ebooks.[35] In 2014, Tom Kabinet launched a secondhand ebook store that allowed readers to resell their ebooks after certifying that they were legally purchased and that any local copies had been deleted. Tom Kabinet was promptly sued by the Dutch Publishers Association. After initially refusing to shut down the site in light of the UsedSoft decision, the court ruled that the site could remain open only if it took additional steps to ensure that the ebooks it resold were lawfully acquired.[36] In short order, Tom Kabinet implemented a digital watermark system. Although not a guarantee against infringement, the Dutch Court of Appeals (Hof Amsterdam) refused to shutter Tom Kabinet's resale marketplace.[37]

Both our proposed exhaustion framework and the one being adopted in Europe embrace shifting focus away from particular copies to instead thinking about rights to use particular works. Copyright, as the name implies, has been preoccupied with copies for a long time. But in the digital world, copies are everywhere. Instead of determining whether a particular behavior is lawful by carefully counting copies, we think courts should be focused on tracking who has rights to use and enjoy a work. A more radical—but admittedly more elegant—way to achieve this result is by taking the copy out of copyright altogether. Christina Mulligan has suggested we solve the digital exhaustion problem, among others, by eliminating copyright law's reproduction right and hinging liability on commercially valuable uses of a work, like display or performance.[38] If creating a new copy didn't trigger infringement liability, digital goods would stand on the same footing as analog ones. You could transfer them freely. Infringement would occur when a nonowner displayed or performed a work, even privately. This proposal is promising, but its implications go well beyond the question of ownership.

Generally, we prefer reforms that place considerable authority in the hands of the courts, but there is no shortage of ways that policymakers could update copyright law to reinforce ownership in the digital marketplace. But often, legal change—particularly when it comes to intellectual property—is prompted by developing technologies. We've already discussed how companies like ReDigi created software that forces us to rethink both the application of the law and our assumptions about the nature of resale markets. We will now outline other technologies with the potential to change the way we conceptualize ownership.

The Role of Technology

As early as the late 1990s, "forward and delete" technologies were under discussion in copyright policy circles. These software tools—purely theoretical at the time—would have kept users honest when they transferred ownership of digital goods.[39] So if you sold or lent an ebook to a friend, the software would *forward* the file to them and *delete* your local copy. The fact that technology capable of safeguarding those sorts of transfers was unavailable at the time contributed to the failure of Representative Boucher's effort to update copyright exhaustion.[40] It also led to the Copyright Office's 2001 decision to oppose expanding exhaustion to digital goods.[41]

Serious efforts to conceptualize and develop technologies that would allow for transfer of ownership of digital goods didn't begin for nearly a decade. One early effort was led by the Institute of Electrical and Electronics Engineers (IEEE), an association of engineers, scientists, and other technical experts and one of the world's most influential standards-setting bodies. In 2010, the IEEE formed a working group to develop a standard for "consumer-ownable digital personal property" (DPP).[42] But to date, no standard has emerged.

Around the same time, companies like Apple and Amazon were developing systems that would enable transfers of digital content between consumers. Amazon for example patented a "secondary market for digital objects."[43] That system lets a user store ebooks, audio, video, and applications in a cloud locker. When the user decides they no longer want it, the system allows the user to transfer their now-used digital content to another user. At that point, the digital content is deleted from their account. Similarly, Apple patented a method of "managing access to digital content items."[44] Apple's system also permits the transfer of digital content between users. It envisions that when a user sells their purchased digital content to a second user, an online store—here, iTunes—stores data about these transactions to establish "which user currently has access to the digital content." Once the user transfers their music, movie, or other purchase, they are "prevented from accessing the digital content." Other patents on "providing a market for digital goods" and a "secondary marketplace for digital media content" cover much the same territory.[45]

No major U.S. retailer has deployed these technologies yet, although there are shades of this approach in Amazon's experiments with restricted ebook "lending." Nonetheless, these patents suggest two things. First, these technologies have moved from theoretical speculation to practical reality. Second, the dominant players in the digital retail market recognize the economic potential of systems that facilitate property-like rights in digital assets. But the patents also reveal that those systems, at least as envisioned by Amazon and Apple, incorporate some potentially troubling limitations. Amazon, for example, describes suspending or terminating transfers of digital content after an unspecified number of transactions, presumably defined in a license agreement and bowing to the demands of publishers.

Apple's patent contemplates a different kind of burden on transfers—a portion of the resale price would be diverted to the publisher.[46] Resale royalties, which give the original creator of a product an ongoing cut of future sales, have been the subject of debate in the courts and Congress for decades. Bills that would create this right—reminiscent of J. K. Rowling's goblin property discussed in chapter 2—have been repeatedly introduced and defeated in the muggle Congress, most recently with the American Royalties Too (ART) Act of 2014.[47] And California's state-level Resale Royalty Act was recently ruled unconstitutional as to sales that take place outside of that state's borders.[48] Typically, this legislation focuses on visual artists—painters, sculptors, and the like—on the grounds that they are especially unlikely to capture the full value of their works at the time of their initial sale. Because resale royalties tend to favor only the most successful visual artists and introduce costly bureaucracies, the case in favor of them is weak.[49] When applied to record labels and movie studios—hardly the victims of unequal bargaining power in the market—that case is nonexistent.

These two aspects of the Apple and Amazon patents point to a more general problem with privately administered digital markets. When markets are run according to rules negotiated between copyright owners and technology platform providers, we trade property rights for conditional privileges. Private actors should not be in a position to define what owners can do with their property, even if they write the next generation of license agreements in a more consumer-friendly way. The baseline for property rights should be a function of the law, not contingent on the kindness of copyright holders and retailers. So these ersatz digital markets might give the appearance of property rights, but what they would actually provide is a slightly relaxed set of license restrictions.

That might be better than nothing, but we don't think it's good enough. There's no shortage of reasons to favor the kind of free secondary marketplace that genuine property rights would enable over these tightly controlled sandboxes. For one, privacy and anonymity would be sacrificed. Retailers and publishers would know what books you bought, sold, and borrowed, not to mention who you borrowed them from or lent them to. And because Apple, Amazon, and other retailers are likely to operate distinct, non-interoperable platforms, these systems balkanize the marketplace. They also increase the risk of lock-in. If your ebook collection and network of lenders, borrowers, and resellers is within the Apple ecosystem, you're far less likely to make the switch to Kindle, for example.

What's more, since their rights would hinge on the deals reached between each platform and publisher, consumers would have to contend with significant information costs. Some digital content could be lent, but not resold. Some could be lent two times in the course of a month, but not three. And some couldn't be lent or resold at all. Consumers would be expected to determine what set of rules apply for each ebook, movie, or album they purchase. Avoiding these information costs is one of the chief reasons we have clear property rules.

We deserve better. There are two primary ways to improve on these sorts of systems. The first is to make them platform neutral. Regardless of who administers the technology, transfers should not be confined to a single provider's ecosystem. If an Apple user wants to lend a digital movie to an Amazon customer, technology should not stand in their way. We could achieve neutrality by clearing the way for third-party technologies that have no economic reason to favor one platform over another. The second way to improve these technology-driven solutions is to provide owners with clear rules about what transfers are allowed. Those rules should be defined by publicly made law, not by private licenses. But even with those changes in place, these solutions rely fundamentally on DRM to police the behavior of owners. These technologies are meant to "keep consumers honest" by taking decisions out of their hands and letting software code make them instead. Even when it is designed to facilitate exhaustion, we find DRM troubling for the many reasons we've already outlined. If DRM was the only way to create a workable digital exhaustion regime, we might grudgingly accept it. But there's another path forward.

Forward-and-delete DRM tries to ease fears about infringement by controlling how many copies exist before and after a transfer. That approach to the problem, much like the *ReDigi* decision's painstaking counting of copies, is built on twentieth-century thinking. It assumes that copies are valuable, long-lasting, and hard to come by. But today, because of the basic architecture of our information networks, copies—lawful and unlawful—are everywhere. What should concern rights holders and policymakers isn't who has a copy of a work, but who has the right to use and transfer it. Reliable evidence about who owns those rights would go a long way toward easing the transition to digital exhaustion.

In a sense, the problem facing lawmakers when it comes to transferring digital goods is one every property system has to confront. Namely, how do we verify ownership to prevent invalid transfers? For particularly valuable assets, we rely on elaborate and costly systems of documentation. Your house has a deed, and your car has a title. Both are registered in centralized public records. These records establish title. They provide legally meaningful answers to the question of ownership, and they help potential purchasers confirm that they are dealing with the right seller. We do something similar for IP rights; copyrights, patents, and trademarks are recorded in searchable databases by the relevant authorities.

With small items of personal property, we typically rely on possession to establish ownership. If your phone is in your pocket or your watch is on your wrist, we assume it's yours. This less formal system makes sense for two reasons. First, it would be far too expensive to keep exhaustive title records for every pair of socks or can of beans that gets sold. And second, most of the time only one person is in possession of a particular tangible object. The object and the ability to exploit its value are deeply intertwined. But neither of these established approaches works all that well for digital assets. Your mp3 collection isn't valuable enough to warrant an official system of recordation. And because digital files are trivial to reproduce, possession in itself tells us very little about legal entitlement.

Surprisingly, cryptocurrencies like bitcoin may help solve the problem of tracking rights in digital assets.[50] Bitcoin is a payment system and corresponding digital currency created in 2008. It is not governed by any central authority; there is no government, central bank, or financial institution standing behind the over $3 billion of bitcoin in the market. Instead, bitcoin relies on its core underlying innovation—the block chain—to verify transactions. Fundamentally, the block chain is a record of transactions. It functions much like the title records at your local county clerk's office or your own checkbook ledger—except the block chain keeps track of every single bitcoin transaction across the globe, updated every ten minutes, to provide a complete and reliable record of ownership. The block chain is not a miracle cure.[51] But it may provide some insight into how to create a workable system of digital personal property.

What sets the block chain apart from other recordation systems is that it's publicly maintained. Unlike your checkbook or title records at the DMV, there is no centralized authority that maintains the block chain. It's the result of an ingeniously complex, cooperative effort. That means the block chain costs very little to maintain, but is highly resistant to manipulation. Trust is essential; if users can't rely on the information it provides, a ledger like the block chain has no value.

While bitcoin remains a large-scale experiment in digital currency, the underlying technology is application-neutral. As Marc Andreessen, whose venture capital firm has invested $50 million in bitcoin-related companies, wrote in the *New York Times*: "Bitcoin gives us, for the first time, a way for one Internet user to transfer a unique piece of digital property to another Internet user, such that the transfer is guaranteed to be safe and secure, everyone knows that the transfer has taken place, and nobody can challenge the legitimacy of the transfer. The consequences of this breakthrough are hard to overstate."[52]

So how does the block chain achieve this seeming miracle? Let's start by thinking about an individual transfer and see how it is verified and recorded. Today, the primary application for the block chain is tracking ownership of bitcoins, but ownership interests in any asset—digital or tangible—could be tracked in the same way. Imagine you are buying a used ebook on a market that implements block chain technology. First, you would want to be sure that the payment you send and the ebook you want to receive can't be intercepted by a malicious third party. You can avoid that by using what's called public key cryptography. This basic approach has been used since the 1970s and forms the basis for popular email encryption programs like Pretty Good Privacy (PGP).

Encryption protects you against third parties, but how do you know you can trust the seller? The seller may not actually own the asset that they have promised to sell. Or maybe they have already promised to sell it to someone else. This problem is familiar from the world of tangible property. It's why you do a title search before buying a house. And it's why eBay has a reputation system. But the problem is even more challenging for digital assets. Unlike the seller of a rare vinyl record or a suburban split-level, the owner of an ebook can just make a second copy with a stroke of the keyboard. How do we prevent them from trying to sell that single asset to two unknowing buyers? And if they do, how do we decide who is the rightful owner?

That's where the public ledger comes in. As a comprehensive and up-to-date record of transactions, it allows anyone to verify transfers of ownership and catch fraud before it happens.[53] So when you go to buy your ebook, you—or more likely, some software on your device—would check whether the seller actually owns it. If they already sold it to someone else or never owned it in the first place, that would be reflected in the ledger, and the transaction would be canceled.

But if all goes well, you pay for the ebook, and your purchase is entered into the ledger where it will be bundled together with a number of other transactions that make up a "block." Bitcoin, for example, bundles transactions into a new block every ten minutes, but that time period can be adjusted as needed. The sequential addition of new blocks is what forms the block chain. Once a block is added to the chain, it becomes part of a complete record of transactions that track changes in the ownership of every asset in the system.

Safeguarding the accuracy of the block chain is obviously of major importance. Someone with the ability to add false information to the chain has the power to reassign ownership of digital assets. Normally, we manage this risk by centralizing control in the hands of some trusted official—the county clerk or the Patent and Trademark Office. But the block chain has no central authority. Instead, the system is trustworthy because adding a block requires a significant investment of resources. That fact doesn't completely prevent false information making its way into the block chain. But it does provide a high enough degree of reliability for users to treat the block chain as proof of title.

Blocks are added to the chain through a process called mining. Miners use computers—in many cases, machines built specifically for the task—to solve a block. Without overwhelming you with technical detail, solving a block involves a sophisticated guessing game. And the more miners there are, the harder it is to guess the right answer. Whoever wins this mathematical lottery gets to add the block to the chain and receives some modest financial reward. With tens of thousands of miners competing to verify transactions and add them to the block chain, a would-be crook would need to consistently guess correctly faster than the rest of the mining community, a task that would take a nearly impossible degree of computing power. And the larger the network of miners grows, the more secure and valuable the block chain ledger becomes.

Relying on the block chain technology pioneered by bitcoin, we can envision a marketplace for digital assets. In that marketplace, consumers could buy, sell, lend, and trade the ebooks, music, movies, applications, and games they buy—and even virtual objects they discover or craft, like the Jade Rabbit, a powerful weapon in the video game *Destiny*.[54] Those transactions would be secure and verifiable, guarding against cheating that could harm both consumers and IP rights holders. The public ledger promises the technological infrastructure to help us transition to what legal scholar Joshua Fairfield calls "bitproperty"—property interests that are decoupled from any tangible object.[55]

Conclusion

Everyday objects are being replaced or supplemented by information. The media we consume is stored in the cloud, not in our hands. Our cars, watches, and clothes—though still physical—incorporate a layer of code that both increases and constrains their functionality. A digital economy structured around interconnected devices and data holds immense promise.

But it also entails risks. Perhaps most troublingly, this new economy has the power to redefine or even eliminate the notion of personal property. If we aren't careful, ownership will become a thing of the past. The loss of ownership puts us all at risk of exploitation. It imposes significant but broadly dispersed costs on society. And it takes decisions about how to live our lives out of our hands and entrusts those choices to a handful of private companies.

Technology alone, no matter how groundbreaking, can't fix the ownership problem. But in conjunction with meaningful legal reform, new innovations can preserve the notion of personal property in this emerging economy. Without legal change, those same technologies become just another tool for rights holders to enforce restrictions on our behavior. Code can reinforce property rights; it can make them easier to transfer and cheaper to track. But it can't create them. Ultimately, property rights are a product of law. Calling an interest a property right—whether it's a neighbor's interest in their home, a copyright holder's interest in their expression, or your interest in the products you buy—is a statement about the degree to which the law will protect that interest in the face of competing claims.[56]

The label "property" carries a great deal of rhetorical force. That's why patent and copyright holders have adopted the language of property, and why they have seen such success in both the courts and Congress in their efforts to strengthen, expand, and extend those rights. But those efforts themselves reveal something crucial about property as an institution. What counts as property, the specific rules and exceptions, and the way we resolve conflict between property owners, are things that change over time. They are decisions that we—through the legislative and judicial processes—make in response to changing conditions and values. Once lawmakers realize that both IP rights holders and consumers can lay equal claim to the property mantle, they are better positioned to balance their competing, and in some ways complementary, interests.

We think meaningful personal property rights in digital assets will benefit consumers, creators, and the market as a whole. Consumers get stable and predictable access, greater privacy protections, and the freedom to make economically and socially valuable uses of the products they purchase. That added value provides something essential from the perspective of creators—a good reason for people to spend their money. People want the freedom to lend, resell, and give away the things they own, and they are willing to spend more to get it. And because ownership reduces information costs and increases competition, it increases the efficiency of the market overall.

None of that is to say that ownership is the only model we need in the digital economy. What we stress is the importance of meaningful choice. We should all have the opportunity to pick and choose from a menu of options, just as we always have. Sometimes rentals make sense, sometimes subscriptions, and sometimes ownership. But a market overrun with complex bespoke bundles of rights like our current licensing-driven regime defeats the purpose of property. It offers no clarity, no certainty, and imposes massive costs on the public. It survives today as the result of misinformation, the absence of genuine alternatives, and legal inertia. Embracing digital ownership would address all three of those problems. The ideal set of rights in the digital property bundle will almost certainly be different from its analog counterpart. And calibrating those rights will not be an easy task. But unless we are content to read about personal property rights in history books—ones that we can't lend and don't own—crafting a notion of ownership applicable to digital goods is a task we can't put off any longer.

Notes

1 Introduction

1. Brad Stone, "Amazon Erases Orwell Books from Kindle," *New York Times*, July 17, 2009, http://www.nytimes.com/2009/07/18/technology/companies/18amazon.html, accessed June 14, 2015.

2. If you borrowed a hard copy of this book from a friend or the local library, you are benefitting from their property right to lend the book. We discuss libraries in depth in chapter 6.

3. Paul Bentley, "How a Third of Bestselling Ebooks Cost More Than the Same Title in Hardback," *Daily Mail* (London), September 30, 2012, http://www.dailymail.co.uk/sciencetech/article-2211022/How-bestselling-ebooks-cost-MORE-title-hardback.html, accessed June 14, 2015.

4. Mark King, "Amazon Wipes Customer's Kindle and Deletes Account with No Explanation," *Guardian* (UK), October 22, 2012, http://www.theguardian.com/money/2012/oct/22/amazon-wipes-customers-kindle-deletes-account, accessed June 14, 2015.

5. Peter Cohen and Jeff Livingston, "More Than Half of U.S. Public Schools Don't Have Adequate Wireless Access," *Atlantic*, November 13, 2013, http://www.theatlantic.com/education/archive/2013/11/more-than-half-of-us-public-schools-dont-have-adequate-wireless-access/281410/, accessed June 14, 2015

6. Nate Hoffelder, "Scholastic to Close Storia eBookstore; Customers Could Lose Access to Their eBook Purchases," *The Digital Reader* (blog), July 27, 2014, http://the-digital-reader.com/2014/07/27/scholastic-close-storia-ebookstore-customers-will-lose-access-ebook-purchases/#.U_fFdvSE-a5, accessed June 14, 2015.

7. "Kindle Store Terms of Use," Amazon Digital Services, Inc., last modified September 6, 2012, http://www.amazon.com/gp/help/customer/display.html?nodeId=201014950, accessed June 14, 2015.

8. Ibid.

9. "Conditions of Use," Amazon Services LLC, last modified December 5, 2012, http://www.amazon.com/gp/help/customer/display.html/?nodeId=508088, accessed June 14, 2015.

10. See, e.g., Ron Wyden, "Regulatory Hardball about Software," *Wall Street Journal*, October 13, 2015, http://www.wsj.com/articles/regulatory-hardball-about-software-1444776652, accessed November 27, 2015.

11. Julie Bosman, "Publisher Limits Shelf Life for Library E-Books," *New York Times*, March 14, 2011, http://www.nytimes.com/2011/03/15/business/media/15libraries.html, accessed June 14, 2015; Jeremy Greenfield, "What Is Going on with Library E-Book Lending?" *Forbes*, June 22, 2012, http://www.forbes.com/sites/jeremygreenfield/2012/06/22/what-is-going-on-with-library-e-book-lending/, accessed June 14, 2015.

12. Michael Kelley, "Random House Says Libraries Own Their Ebooks," *LJ Insider* (blog), *Library Journal*, October 18, 2012, http://lj.libraryjournal.com/2012/10/opinion/random-house-says-libraries-own-their-ebooks-lj-insider, accessed June 14, 2015.; Peter Brantley, "Random House Did Not Mean Own, Exactly," *PWxyz* (blog), *Publisher's Weekly*, October 23, 2012, https://web.archive.org/web/20150626112010/http://blogs.publishersweekly.com/blogs/PWxyz/2012/10/23/just-another-word/ (site discontinued), accessed June 14, 2015.

13. Matt Buchanan, "Five Stores That Hosed Customers with DRM," *Gizmodo* (blog), April 28, 2008, http://gizmodo.com/384741/five-stores-that-hosed-customers-with-drm, accessed June 14, 2015; Jon Healey, "Yahoo Pulls An MSN Music (Only Faster)," *Bit Player* (blog), *Los Angeles Times*, July 23, 2008, http://opinion.latimes.com/bitplayer/2008/07/yahoo-pulls-and.html, accessed June 14, 2015. Walmart later reversed course: Antone Gonsalves, "Wal-Mart Reverses Decision to Shutdown Digital Music DRM Servers," *InformationWeek*, October 10, 2008, http://www.informationweek.com/wal-mart-reverses-decision-to-shutdown-digital-music-drm-servers-/d/d-id/1072848, accessed June 14, 2015.

14. Julie Jacobson, "Perils of DRM: What Happens to Your Digital Content if the Provider Goes Out of Business?" *CEPro*, June 5, 2009, http://www.cepro.com/article/print/what_happens_to_your_digital_content_if_the_provider_goes_out_of_business/, accessed June 14, 2015.

15. Julie E. Cohen, "The Right to Read Anonymously: A Closer Look at Copyright Management in Cyberspace," *Connecticut Law Review* 28 (Summer 1996): 981–1039, 982; R. Anthony Reese, "The First Sale Doctrine in the Age of Digital Networks," *Boston College Law Review* 44 (March 2003): 577–652, at 584.

16. Henry Miller's novel was banned upon its initial release in 1934. The book was finally held not to be obscene by the Supreme Court three decades later in *Grove Press v. Gerstein*, 378 U.S. 577 (1964).

17. The album by rap group 2 Live Crew was declared obscene months after its release and after 1.7 million copies had been sold. Skyywalker Records, Inc. v. Navarro, 739 F. Supp. 578, 582 (S.D. Fla. 1990), *rev'd sub nom.* Luke Records, Inc. v. Navarro, 960 F.2d 134 (11th Cir. 1992). That determination was later reversed on appeal.

18. Nearly two decades after the release of this Academy Award-winning film, it was deemed obscene by a judge in Oklahoma City who demanded copies be confiscated by local authorities. That decision too was eventually overturned. See Camfield v. City of Okla. City, 248 F.3d 1214, 1217 (10th Cir. 2001).

19. Iain Thompson, "Adobe Spies on Readers: EVERY DRM Page Turn Leaked to Base over SSL," *Register* (London), October 23, 2014, http://www.theregister.co.uk/2014/10/23/adobe_updates_digital_editions_encryption/, accessed June 15, 2015.

20. Glenn Greenwald, *No Place to Hide: Edward Snowden, the NSA, and the U.S. Surveillance State* (New York: Metropolitan Books, 2014).

21. Throughout this book, we have opted to use the gender-neutral singular "they" wherever possible. See Jeff Guo, "Sorry, grammar nerds. The singular 'they' has been declared Word of the Year," *Washington Post*, January 8, 2016, https://www.washingtonpost.com/news/wonk/wp/2016/01/08/donald-trump-may-win-this-years-word-of-the-year/, accessed March 16, 2016.

22. R. H. Coase, "The Problem of Social Cost," *Journal of Law and Economics* 3 (October 1960): 1–44, at 15–28.

23. Thomas W. Merrill and Henry E. Smith, "Optimal Standardization in the Law of Property: The Numerus Clausus Principle," *Yale Law Journal* 110 (October 2000): 1–70, at 40–42.

24. Arthur Pigou, *The Economics of Welfare* (1920; repr., New York: Palgrave Macmillan, 2013), 184–185.

25. For better or worse the 2014 film, *The Interview*, was initially withheld from distribution due to fears over its criticism of North Korea. Drew Harwell and Ellen Nakashima, "Hackers' Threats Prompt Sony Pictures to Shelve Christmas Release of 'The Interview,'" *Washington Post*, December 17, 2014, https://www.washingtonpost.com/business/economy/top-movie-theater-chains-cancel-premiere-showings-of-the-interview/2014/12/17/dd1bdb2a-8608-11e4-9534-f79a23c40e6c_story.html, accessed June 15, 2015. Disney maintains a "vault" of films that are unavailable for home viewing, in part as a means of introducing scarcity and driving demand, but also in part to ensure that films like 1946's *Song of the South* remain unseen if not forgotten. Jason Sperb, *Disney's Most Notorious Film: Race, Convergence, and the Hidden Histories of* Song of the South (Austin: University of Texas Press, 2012).

26. Eric von Hippel, *Democratizing Innovation* (Cambridge, MA: MIT Press, 2005).

27. Sony Corp. of Am. v. Universal City Studios, Inc., 464 U.S. 417, 446 n.27 (1984).

28. See 17 U.S.C. § 504(c) (2012).

29. The Pirate Bay, for those unfamiliar, is a popular site that allows users to contribute and locate torrent files that facilitate the often unauthorized distribution of copyrighted material.

2 Property and the Exhaustion Principle

1. Brad Dashoff and John Antonacci, "Understanding Real Property Interests and Deeds," *ABA GPSolo Law Trends & News* 7, no. 4 (Summer 2011), http://www.americanbar.org/newsletter/publications/law_trends_news_practice_area_e_newsletter_home/2011_summer/real_property_interests_deeds.html, accessed June 15, 2015.

2. For a more detailed exposition of these devices, see Gerald Korngold, *Private Land Use Arrangements: Easements, Real Covenants, and Equitable Servitudes* (Huntington, NY: Juris Publishing, 2004), 287.

3. The courts have imposed some important limitations on the power of real property owners to saddle future generations with restrictions on the use of that property. Racially restrictive covenants, which prohibited the sale of real estate to nonwhites, were deemed unconstitutional in *Shelley v. Kraemer*, 334 U.S. 1 (1948).

4. Christina Mulligan, "A Numerus Clausus Principle for Intellectual Property," *Tennessee Law Review* 80 (Winter 2013): 235–290, at 251–252.

5. Courts have historically proven hostile to servitudes on chattels due to their tendency to provide insufficient notice, restrain alienation, limit future unforeseen usability, and impose high information costs on subsequent purchasers. See Zechariah Chafee Jr., "The Music Goes Round and Round: Equitable Servitudes and Chattels," *Harvard Law Review* 69, no. 7 (May 1956): 1250–1264, at 1261; Molly Shaffer Van Houweling, "The New Servitudes," *Georgetown Law Journal* 96 (2008): 885–950, at 897–898.

6. See, e.g., De Mattos v. Gibson (1858) 4 De G&J 276 (Eng.); Taddy & Co. v. Sterious & Co. (1904) 1 Ch. 354 (Eng.); John D. Park & Sons Co. v. Hartman, 153 F. 24, 39 (6th Cir. 1907); Miles Med. Co. v. John D. Park & Sons Co., 220 U.S. 373 (1911); Van Houweling, "The New Servitudes." But see Glen O. Robinson, "Personal Property Servitudes," *University of Chicago Law Review* 71 (Fall 2004): 1449–1523 (arguing in favor of servitudes on personal property).

7. Christina Mulligan, "Personal Property Servitudes on the Internet of Things," *Georgia Law Review* (forthcoming).

8. John D. Park & Sons Co. v. Hartman, 153 F. 24, 39 (6th Cir. 1907).

9. J. K. Rowling, *Harry Potter and the Deathly Hallows* (New York: Arthur A. Levine Books, 2007), 417–418.

10. Exhaustion also plays a role in trademark law, where it permits the resale of authentic goods without the trademark holder's permission. See generally Yvette Joy Liebesman and Benjamin Wilson, "The Mark of a Resold Good," *George Mason Law Review* 20 (Fall 2012): 157–205.

11. See Feist Publ'ns, Inc. v. Rural Tel. Serv. Co., 499 U.S. 340 (1991).

12. 17 U.S.C. § 102(a) (2012).

13. Ibid., § 106.

14. See 35 U.S.C. § 154(a)(2) (2012); 17 U.S.C. § 302(a) (2012).

15. Thomas Jefferson to Isaac McPherson, 13 August 1813, in *The Writings of Thomas Jefferson*, ed. Andrew A. Lipscomb and Albert Ellery Bergh (Washington, DC: Thomas Jefferson Memorial Association, 1905), 334.

16. Joshua A.T. Fairfield, "Virtual Property," *Boston University Law Review* 85, no. 4 (October 2005): 1047–1102.

17. William Blackstone, *Commentaries on the Laws of England*, ed. John L. Wendell, vol. 2 (New York: Harper & Bros., 1857), 1.

18. See Albert W. Alschuler, "Rediscovering Blackstone," *University of Pennsylvania Law Review* 145 (November 1996): 30–32.

19. John Locke, "Of Property," chap. 5 in *Second Treatise of Government*, ed. Jonathan Bennett (2010), http://www.earlymoderntexts.com/assets/pdfs/locke1689a_1.pdf, accessed June 16, 2015.

20. *Hegel's Philosophy of Right*, trans. T. M. Knox (London: Oxford University Press, 1820).

21. Joseph Campbell, *The Hero with a Thousand Faces* (New York: Pantheon Books, 1949).

22. The Copyright Act exempts certain public performances of audiovisual works from liability so long as, among other things, the diagonal screen size is no greater than fifty-five inches. Seriously. 17 U.S.C. § 110(5)(B) (2012).

23. Unlike the term "publicly" as used on the public performance and display rights, the phrase "to the public" as used in the distribution right is undefined in the Copyright Act. Few cases have considered the implications of that phrase.

24. Thad Moore, "Here's Why Atari Fans Just Spent $100,000 on Video Games from a Dump," *Washington Post*, September 4, 2015, https://www.washingtonpost.com/news/the-switch/wp/2015/09/04/heres-why-atari-fans-just-spent-100000-on-video-games-from-a-dump, accessed November 25, 2015.

25. Ibid., § 109(c), 90 Stat. at 2549 (current version at 17 U.S.C. § 109[c] [2012]).

26. Ibid., at § 117, 90 Stat. at 2565 (current version at 17 U.S.C. § 117 [2012]).

27. Clemens v. Estes, 22 F. 899 (D. Mass. 1885).

28. Harrison v. Maynard, Merrill & Co., 61 F. 689 (2d Cir. 1894).

29. Doan v. Am. Book Co., 105 F. 772 (7th Cir. 1901).

30. Kipling v. G. P. Putnam's Sons, 120 F. 631 (2d Cir. 1903).

31. Bobbs-Merrill Co. v. Straus, 210 U.S. 339 (1908).

32. 1909 Copyright Act: An Act to Amend and Consolidate the Acts Respecting Copyright, Pub. L. No. 60-349, ch. 320, § 41, 35 Stat. 1075, 1084 (1909).

33. Copyright Act of 1976, Pub. L. No. 94-553, § 109, 90 Stat. 2541, 2548–49 (1976) (current version at 17 U.S.C. § 109 [2012]).

34. Kirtsaeng v. John Wiley & Sons, Inc., 133 S. Ct. 1351 (2013).

35. There are half measures between the full international exhaustion regime the Court adopted in *Kirtsaeng* and the place-of-manufacture rule endorsed by *Wiley*. Some countries have adopted regional exhaustion where a sale in one country exhausts rights in a defined number of neighboring territories, but not the world at large. Ariel Katz, "The First Sale Doctrine and the Economics of Post-Sale Restraints," *Brigham Young University Law Review* 2014, no. 1 (2014): 57–142, at 75. Another option suggested in Justice Kagan's concurring opinion in *Kirtsaeng* would preclude unauthorized importation of a lawfully sold copy, but once it reached the United States would recognize the first sale doctrine. *Kirtsaeng*, 133 S. Ct. at 1372–1373 (Kagan, J., concurring).

36. Katz, "The First Sale Doctrine," 82.

37. Leah Shaver, "Copyright and Inequality," *Washington University Law Review* 92 (2014): 132 ("When asked why they do not read more often, respondents overwhelmingly cite the affordability and availability of books as the primary barriers. The most common answers include: 'Books are expensive' (45 percent), 'Books are so expensive that you cannot afford to buy them' (33 percent).").

38. Shamnad Basheer et al., "Exhausting Copyrights and Promoting Access to Education: An Empirical Take," *Journal of Intellectual Property Rights* 17 (2012): 335–344, at 340.

39. Carolyn Bigda, "Money-saving Tips for College Students," *Chicago Tribune*, August 11, 2015, http://www.chicagotribune.com/business/yourmoney/sc-cons-0813-started-20150810-column.html, accessed November 25, 2015.

40. Ben Popken, "College Textbook Prices Have Risen 1,041 Percent Since 1977," *NBC News*, August 6, 2015, http://www.nbcnews.com/feature/freshman-year/college-textbook-prices-have-risen-812-percent-1978-n399926, accessed November 25, 2015.

41. Jonathan Band and Jonathan Gerafi, *Profitability of Copyright-Intensive Industries* (Washington, DC: infojustice.org, 2013), PDF report, http://infojustice.org/wp-content/uploads/2013/06/Profitability-of-Copyright-Industries.pdf, accessed July 6, 2015.

42. Ibid.; Heather Morrison, "Chapter Two: Scholarly Communication in Crisis," in "Freedom for Scholarship in the Internet Age" (doctoral dissertation, Simon Fraser University, 2011), https://web.archive.org/web/20150828020037/http://pages.cmns.sfu.ca/heather-morrison/chapter-two-scholarly-communication-in-crisis/ (site discontinued), accessed July 6, 2015; see also "Of Goats and Headaches," *Economist*, May 26, 2011, http://www.economist.com/node/18744177/, accessed July 6, 2015 (noting Elsevier's 36 percent profit margins).

43. Mark J. Perry, "The Public Thinks the Average Company Makes a 36% Profit Margin, Which Is about 5X Too High," *AEIdeas*, April 2, 2015, https://www.aei.org/publication/the-public-thinks-the-average-company-makes-a-36-profit-margin-which-is-about-5x-too-high, accessed July 6, 2015.

44. It's worth noting that the college textbook market is particularly susceptible to insensitivity to rising prices since those who make decisions about which books are required in a course, namely professors, don't pay for them.

45. Twentieth Century Music Corp. v. Aiken, 422 U.S. 151 (1975).

46. Fred R. Byers, "Care and Handling of CDs and DVDs—A Guide for Librarians and Archivists" (Gaithersburg, MD: National Institute of Standards and Technology, 2003), http://www.itl.nist.gov/iad/894.05/docs/CDandDVDCareandHandlingGuide.pdf, accessed July 6, 2015; "Preserving CDs and DVDs," National Archives of Australia, http://www.naa.gov.au/records-management/agency/preserve/physical-preservation/CDs-and-DVDs.aspx, accessed November 15, 2015.

47. Chuck Philips, "Compact Disc War Headed for the Court," *Los Angeles Times*, July 31, 1993, http://articles.latimes.com/1993-07-31/entertainment/ca-18861_1_compact-discs, accessed November 15, 2015.

48. Ibid.

49. James Lardner, *Fast Forward: Hollywood, the Japanese and the VCR Wars* (New York: W. W. Norton, 1987), 179.

50. Joshua M. Greenberg, *From Betamax to Blockbuster: Video Stores and the Invention of Movies on Video* (Cambridge, MA: MIT Press, 2010), 119.

51. "Terms of Use," Beachbody, LLC, last modified July 18, 2014, https://www.beachbody.com/product/about_us/terms_of_use.do, accessed July 6, 2015.

52. See 17 U.S.C. § 109(b) (2012).

53. Ellen Forman, "Nintendo Zaps Blockbuster Reproduction of Game Instructions Spurs Copyright Lawsuit," *Sun Sentinel* (Fort Lauderdale), August 13, 1989, http://articles.sun-sentinel.com/1989-08-13/business/8902250572_1_nintendo-blockbuster-video-games, accessed July 6, 2015; Henry Gilbert, "Lawsuits That Altered the Course of Gaming History," *GamesRadar*, February 4, 2014, http://www.gamesradar.com/lawsuits-changed-gaming/, accessed July 6, 2015.

54. Robert Purchese, "Lionhead: Pre-owned Worse Than Piracy," *Eurogamer*, May 17, 2011, http://www.eurogamer.net/articles/2011-05-17-lionhead-pre-owned-worse-than-pc-piracy, accessed July 6, 2015.

55. John Gaudiosi, "GameStop President Tony Bartel Talks Xbox One, PlayStation 4, Used Games and Pre-Orders," *Forbes*, May 21, 2013, http://www.forbes.com/sites/johngaudiosi/2013/05/21/gamestop-president-tony-bartel-talks-xbox-one-playstation-4-used-games-and-pre-orders/, accessed July 6, 2015.

56. Hidehiro Inooka, "Electronic Content Processing System, Electronic Content Processing Method, Package of Electronic Content, and Use Permission Apparatus," US Patent 9,183,358, filed September 12, 2012, issued November 10, 2015, http://www.google.com/patents/US20130007892, accessed November 15, 2015.

57. Timothy Geigner, "Microsoft Capitulates, Removes Online DRM from Xbox One," *Techdirt*, June 19, 2013, https://www.techdirt.com/articles/20130619/13581923535/microsoft-capitulates-removes-online-drm-xbox-one.shtml, accessed July 6, 2015.

58. Wesley Yin-Poole, "Fallout: New Vegas Dev Hopes Digital Distribution 'Stabs Used Game Market in the Heart,'" *Eurogamer*, December 13, 2011, http://www.eurogamer.net/articles/2011-12-13-fallout-new-vegas-dev-hopes-digital-distribution-stabs-used-game-market-in-the-heart, accessed July 6, 2015.

59. See Daniel Gross, "Does a Free Download Equal a Lost Sale," *New York Times*, November 21, 2004, http://www.nytimes.com/2004/11/21/business/yourmoney/does-a-free-download-equal-a-lost-sale.html, accessed July 6, 2015.

60. Jordan Kahn, "Eddy Cue: Apple Passed 35 Billions Songs Sold on iTunes Last Week, 40 Million iTunes Radio Listeners," *9to5Mac*, May 28, 2014, http://9to5mac.com/2014/05/28/eddy-cue-apple-passed-35-billions-songs-sold-on-itunes-last-week-40-million-itunes-radio-listeners/, accessed March 15, 2016.

61. Universal City Studios, Inc. v. Sony Corp. of Am., 480 F. Supp. 429, 467 (C.D. Cal. 1979).

3 Copies, Clouds, and Streams

1. In China, books were printed as early as the late sixth century. And around 1050, Pi Sheng developed movable type, hundreds of years before Gutenberg. John Scales Avery, *Information Theory and Evolution* (Singapore: World Scientific Publishing, 2012), 138–139.

2. For book publishers, the desire to control the press was rooted in economic concerns. Without competition, they could make a tidy sum. But for governments, control over the printing press was a powerful tool for suppressing criticism. Elizabeth L. Eisenstein, *The Printing Press as an Agent of Change* (Cambridge, UK: Cambridge University Press, 1980).

3. For more on the history of Venetian and English printing privileges, see Maurizio Borghi, "A Venetian Experiment on Perpetual Copyright," in *Privilege and Property: Essays on the History of Copyright*, ed. Ronan Deazley, Martin Kretschmer, and Lionel Bentley (Cambridge, UK: Open Book Publishers, 2010), 137–155; Mark Rose, *Authors and Owners: The Invention of Copyright* (Cambridge, MA: Harvard University Press, 1995).

4. Rose, *Authors and Owners*, 14.

5. Prior to the Copyright Act of 1909, authors could generally control public performances of unpublished works. See Ferris v. Frohman, 223 U.S. 424 (1912) (applying the Copyright Act of 1891).

6. Pope v. Curl, 2 Atk. 342, 26 Eng. Rep. 608 (1741).

7. Stephens v. Cady, 55 U.S. 528 (1853).

8. See, e.g., Parton v. Prang, 18 F. Cas. 1273 (C.C.D. Mass. 1872) (No. 10,784); Pushman v. New York Graphic Society, 39 N.E.2d 249 (N.Y. 1942).

9. Copyright Act of 1976, at § 202, 90 Stat. at 2568–69 (current version at 17 U.S.C. § 202 [2012]).

10. "iTunes Music," Apple, Inc., http://www.apple.com/itunes/music/, accessed March 13, 2016 (noting a library of forty-three million songs); Jordan Kahn, "Eddy Cue: Apple Passed 35 Billion Songs Sold on iTunes Last Week, 40 Million iTunes Radio Listeners," *9to5Mac*, May 28, 2014, http://9to5mac.com/2014/05/28/eddy-cue-apple-passed-35-billions-songs-sold-on-itunes-last-week-40-million-itunes-radio-listeners/, accessed March 13, 2016; "iTunes Store Top Music Retailer in the US," Apple Press Info, April 3, 2008, https://www.apple.com/pr/library/2008/04/03iTunes-Store-Top-Music-Retailer-in-the-US.html, accessed March 13, 2016 (noting that iTunes "became the largest retailer in the US based on the amount of music sold during January and February 2008").

11. Jim Milliot, "BEA 2013: The E-book Boom Years," *Publisher's Weekly*, May 29, 2013, http://www.publishersweekly.com/pw/by-topic/industry-news/bea/article/57390-bea-2013-the-e-book-boom-years.html, accessed June 15, 2015; Marisa Bluestone, "U.S. Publishing Industry's Annual Survey Reveals $28 Billion in Revenue in 2014," *Publishers.org*, June 10, 2015, http://publishers.org/news/us-publishing-industry's-annual-survey-reveals-28-billion-revenue-2014, accessed June 15, 2015.

12. Alexandra Alter, "The Plot Twist: E-Book Sales Slip, and Print Is Far from Dead," *New York Times*, September 22, 2015, http://www.nytimes.com/2015/09/23/business/media/the-plot-twist-e-book-sales-slip-and-print-is-far-from-dead.html, accessed November 18, 2015.

13. Capitol Records, LLC v. ReDigi Inc., 934 F. Supp. 2d 640 (S.D.N.Y. 2013).

14. The Supreme Court of Canada came to a different conclusion in a case involving the transfer of images from posters to canvases without the copyright holder's consent. Théberge v. Galerie d'Art du Petit Champlain Inc., 2002 SCC 34 [2002], 2 S.C.R. 336 (Can.). The Court determined that image was "was not reproduced" but "transferred from one display to another" (338).

15. See Computer Software Copyright Act of 1980, Pub. L. No. 96-517, 94 Stat. 3015, 3028–29 (1980) (current version at 17 U.S.C. § 117 [2012]).

16. "Term and Conditions," Apple, Inc., last modified October 21, 2015, http://www.apple.com/legal/internet-services/itunes/us/terms.html, accessed November 18, 2015.

17. Or not. See United States v. Wise, 550 F.2d 1180 (9th Cir. 1977).

18. 134 S. Ct. 2498, 2508 (2014).

19. See White-Smith Music Publ'g. Co. v. Apollo Co., 209 U.S. 1 (1908).

20. Cartoon Network LP, LLLP v. CSC Holdings, Inc., 536 F.3d 121 (2d Cir. 2008).

21. MAI Sys. Corp. v. Peak Computer, Inc., 991 F.2d 511, 519 (9th Cir. 1993).

22. Todd Spangler, "Netflix Tops 57 Million Subscribers in Q4 as U.S. Growth Slows," *Variety*, January 20, 2015, http://variety.com/2015/digital/news/netflix-tops-57-million-subscribers-in-q4-as-u-s-growth-slows-1201409712/, accessed June 15, 2015; "Global Internet Phenomena Report," Sandvine, Inc., https://www.sandvine.com/trends/global-internet-phenomena/, accessed June 15, 2015.

23. "20 Million Reasons to Say Thanks," *Spotify News*, June 10, 2015, https://news.spotify.com/us/2015/06/10/20-million-reasons-to-say-thanks/, accessed June 15, 2015.

24. Ibid.

25. Brent Lang, "Digital Home Entertainment to Exceed Physical by 2016, Study Finds," *Variety*, June 3, 2014, http://variety.com/2014/digital/news/digital-home-entertainment-to-exceed-physical-by-2016-study-finds-1201207708/, accessed November 18, 2015.

26. Ethan Smith, "Music Downloads Plummet in U.S., but Sales of Vinyl Records and Streaming Surge," *Wall Street Journal*, January 1, 2015, http://www.wsj.com/articles/music-downloads-plummet-in-u-s-but-sales-of-vinyl-records-and-streaming-surge-1420092579, accessed June 15, 2015.

27. "Introducing Apple Music—All the Ways You Love Music. All in One Place," Apple Press Info, June 8, 2015, https://www.apple.com/pr/library/2015/06/08Introducing-Apple-Music-All-The-Ways-You-Love-Music-All-in-One-Place-.html, accessed November 18, 2015.

28. Michael Liedtke, "Gaps in Netflix's Online Library Likely to Persist," *Yahoo News*, April 9, 2012, http://news.yahoo.com/gaps-netflixs-online-library-likely-persist-200620994.html; "Information," Spotify Ltd., https://press.spotify.com/us/information/, accessed June 15, 2015; Piotr Kowalczyk, "Kindle Unlimited Ebook Subscription—9 Things to Know," *Ebook Friendly*, last modified October 27, 2015, http://ebookfriendly.com/kindle-unlimited-ebook-subscription/, accessed November 18, 2015.

29. Ken Auletta, "Outside the Box," *New Yorker*, February 3, 2014, http://www.newyorker.com/magazine/2014/02/03/outside-the-box-2, accessed June 15, 2015.

30. When film distributor The Weinstein Company signed an exclusive distribution agreement with Blockbuster, Netflix relied on retail purchases and the first sale doctrine to make those titles available to its customers. See Statement of Reed Hastings, Netflix CEO, "Netflix, Inc. Q3 2009 Earnings Call," October 22, 2009, transcript, http://seekingalpha.com/article/168407-netflix-inc-q3-2009-earnings-call-transcript (noting that because "retail is so big and diffused," studios have been unable to prevent retail acquisition and that because of "liquidity in the supply chain," the purchase of used retail copies saved Netflix money).

31. "Spotify Explained: How Does Spotify Make Money?" Spotify Ltd., http://www.spotifyartists.com/spotify-explained/#how-does-spotify-make-money, accessed June 15, 2015.

32. These other sources of income, particularly live performance, are of increasing importance. Swift's *Red* tour, which wrapped up in 2014, grossed $150 million, more than the next highest-grossing tours from Beyonce, Jay-Z, and Justin Timberlake combined.

33. Taylor Swift, "For Taylor Swift, the Future of Music Is a Love Story," *Wall Street Journal*, July 7, 2014, http://www.wsj.com/articles/for-taylor-swift-the-future-of-music-is-a-love-story-1404763219, accessed June 15, 2015.

34. Daniel Kahneman, Jack L. Knetsch, and Richard H. Thaler, "Experimental Tests of the Endowment Effect and the Coase Theorem," *Journal of Political Economy* 98, no. 6 (1990): 1325–1348.

35. Carey K. Morewedge et al., "Bad Riddance or Good Rubbish? Ownership and Not Loss Aversion Causes the Endowment Effect," *Journal of Experimental Psychology* 45, no. 4 (July 2009): 947–951.

36. Yannick Ferreira De Sousa and Alistair Munro, "Truck, Barter, and Exchange versus the Endowment Effect: Virtual Field Experiments in an Online Game Environment," *Journal of Economic Psychology* 33, no. 3 (June 2012): 482–493. Although this experiment found that the endowment effect was reduced among experienced players of an online role-playing game, that finding is consistent with experiments in the offline world.

37. Keith Caufield, "Vinyl Album Sales Hit Historic High in 2014, Again," *Billboard,* December 31, 2014, http://www.billboard.com/articles/columns/chart-beat/6422442/vinyl-album-sales-hit-historic-high-2014, accessed June 15, 2015.

38. Joshua P. Friedlander, *News and Notes on 2015 Mid-Year RIAA Shipment and Revenue Statistics* (Washington, DC: RIAA, 2015), PDF report, https://www.riaa.com/wp-content/uploads/2015/09/2015_RIAAMidYear_ShipmentData.pdf, accessed November 18, 2015.

39. Jon Fingas, "'The Simpsons' Seasons Won't Be Available on Disc from Now On," *Engadget,* April 12, 2015, http://www.engadget.com/2015/04/12/the-simpsons-drops-disc-releases/, accessed November 18, 2015.

40. Kevin Smith, "Planning for Musical Obsolescence," *Scholarly Communications @ Duke* (blog), Duke University Libraries, July 28, 2014, http://blogs.library.duke.edu/scholcomm/2014/07/28/planning-for-musical-obsolescence/, accessed June 15, 2015.

41. Cory Doctorow, "Oxford English Dictionary—the Future," *Guardian* (UK), August 23, 2013, http://www.theguardian.com/technology/2013/aug/23/oxford-english-dictionary-future-digitally, accessed June 15, 2015.

42. Stephen Shankland, "Adobe Launches Creative Cloud Subscription Service," *CNET,* May 11, 2012, http://www.cnet.com/news/adobe-launches-creative-cloud-subscription-service/, accessed June 15, 2015.

43. Jamie Condliffe, "Microsoft Still Sells Its OS in Boxes—This Is How Windows 10 Will Look," *Gizmodo* (blog), July 13, 2015, http://gizmodo.com/microsoft-still-sells-its-os-in-boxes-this-is-how-windo-1717434336, accessed November 18, 2015.

4 Ownership and the Fine Print

1. See "Licensed Application End User License Agreement," Apple, Inc., http://www.apple.com/legal/internet-services/itunes/appstore/dev/stdeula/, accessed June 15, 2015; "iTunes Terms and Conditions," Apple Inc., last modified June 30, 2015, http://www.apple.com/legal/internet-services/itunes/us/terms.html, accessed July 7, 2015.

2. Tom Gardner, "To Read, Or Not to Read ... the Terms and Conditions," *Daily Mail* (London), March 22, 2012, http://www.dailymail.co.uk/news/article-2118688/PayPal-agreement-longer-Hamlet-iTunes-beats-Macbeth.html, accessed July 7, 2015.

3. Bob Dorman, "Adobe Demands 7,000 Years a Day from Humankind," *Register*, December 4, 2012, http://www.theregister.co.uk/2012/12/04/feature_tech_licences_are_daft, accessed July 7, 2015.

4. Alex Hern, "I Read All the Small Print on the Internet and It Made Me Want to Die," *Guardian* (UK), June 15, 2015, http://www.theguardian.com/technology/2015/jun/15/i-read-all-the-small-print-on-the-internet, accessed July 7, 2015.

5. Douglas E. Phillips, *The Software License Unveiled: How Legislation by License Controls Software Access* (New York: Oxford University Press, 2009), 79.

6. Yannis Bakos, Florencia Marotta-Wurgler, and David R. Trossen, "Does Anyone Read the Fine Print? Consumer Attention to Standard Form Contracts," New York University Law and Economics Working Papers, Paper 195, New York University School of Law, New York, 2014, 22, http://lsr.nellco.org/cgi/viewcontent.cgi?article=1199&context=nyu_lewp, accessed July 7, 2015.

7. Mike Masnick, "Supreme Court Chief Justice Admits He Doesn't Read Online EULAs or Other 'Fine Print,'" *Techdirt*, October 22, 2010, https://www.techdirt.com/articles/20101021/02145811519.shtml, accessed July 7, 2015.

8. Mike Masnick, "Proof That (Almost) No One Reads End User License Agreements," *Techdirt*, February 23, 2005, https://www.techdirt.com/articles/20050223/1745244.shtml, accessed July 7, 2015.

9. George A. Akerlof, "The Market for 'Lemons': Quality Uncertainty and the Market Mechanism," *Quarterly Journal of Economics* 84, no. 3 (August 1970): 488–500, http://www.jstor.org/stable/1879431, accessed July 7, 2015.

10. Services like Carfax have addressed this problem by making information about specific vehicles more readily available. Steven Mufson and Michael A. Fletcher, "Carfax Figures Indicate an 'Alarming Number' of Recalled Cars Are Sold without Being Fixed," *Washington Post*, April 4, 2014, https://www.washingtonpost.com/business/economy/carfax-figures-indicate-an-alarming-number-of-recalled-cars-are-sold-without-being-fixed/2014/04/03/093e9464-bb47-11e3-9c3c-311301e2167d_story.html, accessed March 13, 2016.

11. Marshall Kirkpatrick, "Does Google Have Rights to Everything You Send through Chrome?" *Readwrite*, September 3, 2008, http://readwrite.com/2008/09/03/does_google_have_rights_to_all, accessed July 7, 2015.

12. Nate Anderson, "Google on Chrome EULA Controversy: Our Bad, We'll Change It," *Ars Technica*, September 3, 2008, http://arstechnica.com/tech-policy/2008/09/google-on-chrome-eula-controversy-our-bad-well-change-it/, accessed July 7, 2015.

13. Nancy S. Kim, *Wrap Contracts: Foundations and Ramifications* (Oxford, UK: Oxford University Press, 2013), 60.

14. Ibid.

15. Matthew Humphries, "Retail Copies of Office 2013 Are Tied to a Single Computer Forever," Geek.com, February 13, 2013, http://www.geek.com/news/retail-copies-of-office-2013-are-tied-to-a-single-computer-forever-1539563/, accessed July 7, 2015.

16. For an example of terms that limit liability, see "SoundCloud Terms of Use," SoundCloud Limited, last modified March 12, 2013, https://soundcloud.com/terms-of-use, accessed July 7, 2015. For an example of terms that contain clauses mandating arbitration and barring class actions, see "Dropbox Terms of Service," Dropbox, Inc., last modified January 22, 2015, https://www.dropbox.com/terms2014, accessed July 7, 2015. See also Jessica Silver-Greenberg and Robert Gebeloff, "Arbitration Everywhere, Stacking the Deck of Justice," *New York Times*, October 31, 2015, http://www.nytimes.com/2015/11/01/business/dealbook/arbitration-everywhere-stacking-the-deck-of-justice.html, accessed November 29, 2015.

17. Kim, *Wrap Contracts*, 66–67.

18. "Licensed Application EULA," Apple Inc. Apple's iTunes Store is somewhat less clear in how it characterizes transactions with consumers. After describing those transactions as "purchases" and noting that "all sales ... are final," Apple insists that consumers agree not to "rent, lease, loan, sell, [or] distribute" their purchases. "Terms and Conditions," Apple Inc., last modified October 21, 2015, http://www.apple.com/legal/internet-services/itunes/us/terms.html, accessed November 29, 2015.

19. "Microsoft Software License Agreement: Microsoft Office 2013 Desktop Application Software" Microsoft Corp., https://products.office.com/en-us/microsoft-software-license-agreement, accessed June 16, 2015.

20. "Kindle Store Terms of Use," Amazon Digital Services, Inc., last modified September 6, 2012, http://www.amazon.com/gp/help/customer/display.html?nodeId=201014950, accessed June 14, 2015; see Masnick, "Proof."

21. "SEN Terms of Service," Sony Network Entertainment Europe Ltd., last modified November 2013, https://www.playstationnetwork.com/en-gb/terms-of-service/, accessed June 15, 2015.

22. F.B.T. Prods. v. Aftermath Records, 621 F.3d 958 (9th Cir. 2010); Eriq Gardner, "Universal Music Settling Big Class Action Lawsuit over Digital Royalties," *Hollywood Reporter*, March 19, 2015, http://www.hollywoodreporter.com/thr-esq/universal-music-settling-big-class-783096, accessed June 14, 2015.

23. Watts S. Humphrey, "Software Unbundling: A Personal Perspective," *IEEE Annals of the History of Computing* 34, no. 1 (2002): 59–63, at 60, doi:10.1109/85.988582.

24. The Copyright Act did not embrace software until the Computer Software Copyright Act of 1980, 94 Stat. at 3028–29. The Supreme Court first opened the door to software patents in 1981. See Diamond v. Diehr, 450 U.S. 175 (1981). But the explosion of software patenting didn't occur until the Federal Circuit's 1998 decision in *State St. Bank and Trust Co. v. Signature Fin. Grp., Inc.*, 149 F.3d 1368 (Fed. Cir. 1998).

25. Video games are one important exception. See Computer Software Rental Amendment Act of 1990, Pub. L. No. 101-650, sec. 802, § 109(b), 104 Stat. 5089, 5134–35 (1990) (current version at 17 U.S.C. 109[b] [2012]).

26. For further commentary on how courts have treated copies placed in computer memory, see Aaron Perzanowski, "Fixing RAM Copies," *Northwestern University Law Review* 104, no. 3 (Summer 2010): 1067–1108, http://papers.ssrn.com/sol3/papers.cfm?abstract_id=1441685, accessed June 14, 2015.

27. Doesn't even have resale value without transfer of right to copy.

28. 17 U.S.C. § 117(a)(1) (2012).

29. Ibid., § 117(c).

30. Ibid., § 117(a)(2).

31. Ibid., § 117(b). Adaptation copies can only be transferred with the permission of the copyright holder.

32. Phillips, *The Software License Unveiled*, xii–xiii.

33. See generally Margaret Jane Radin, *Boilerplate: The Fine Print, Vanishing Rights, and the Rule of Law* (Princeton, NJ: Princeton University Press, 2013).

34. See, e.g., Vault Corp. v. Quaid Software, Ltd., 847 F.2d 255 (5th Cir. 1988); Step-Saver Data Sys. v. Wyse Tech., 939 F.2d 91 (3d Cir. 1991).

35. Mark Lemley, "Terms of Use," *Minnesota Law Review* 91 (December 2006): 459–483, at 465; Kim, *Wrap Contracts*, 128.

36. Melvin Aron Eisenberg, "The Limits of Cognition and the Limits of Contract," *Stanford Law Review* 47 (January 1995): 211–259, at 243–244.

37. ProCD, Inc. v. Zeidenberg, 86 F.3d 1447, 1454 (7th Cir. 1996).

38. Ibid., 1452.

39. Chris Newman, "A License Is Not a 'Contract Not to Sue': Disentangling Property and Contract in the Law of Copyright Licenses," *Iowa Law Review* 98 (March 2013): 1103–1160, at 1141.

40. Eben Moglen, "Enforcing the GNU GPL," GNU.org, September 10, 2001, http://www.gnu.org/philosophy/enforcing-gpl.html, accessed July 7, 2015.

41. See 17 U.S.C. § 107 (2012).

42. Jeremy Gordon, "RZA Says Wu-Tang Clan Offered $5 Million for New Album That's Only Available as One Copy," *Pitchfork*, April 2, 2014, http://pitchfork.com/news/54627-rza-says-wu-tang-clan-offered-5-million-for-new-album-thats-only-available-as-one-copy/, accessed July 7, 2015.

43. Brian W. Carver, "Why License Agreements Do Not Control Copy Ownership: First Sales and Essential Copies," *Berkeley Technology Law Journal* 25 (2010): 1887–1954, at 1899.

44. See U.C.C. § 2-401 (2002). Any reservation of title in goods shipped or delivered to a buyer is treated as reservation of a security interest. Reservation of full title is simply not permitted under the U.C.C. when goods are permanently transferred to a buyer. And upon full payment, the buyer is deemed to hold title to goods.

45. For an example of how the Ninth Circuit has utilized objective facts in determining the presence or absence of a first sale, see United States v. Wise, 550 F.2d 1180 (9th Cir. 1977).

46. See Unordered Merchandise Act, 39 U.S.C. § 3009(b) (2012). Unordered merchandise "may be treated as a gift by the recipient, who shall have the right to retain, use, discard, or dispose of it in any manner he sees fit without any obligation whatsoever to the sender."

47. The *Vernor* test is also inconsistent with the result in *Augusto*. Applied to the promotional CDs, each prong of the *Vernor* test is satisfied: (1) UMG characterized the transaction as a license; (2) it prohibited recipients from transferring the discs to others; and (3) it confined them to personal noncommercial use of the discs.

48. See, e.g., DSC Commc'ns Corp. v. DGI Techs., 81 F.3d 597 (5th Cir. 1996); Krause v. Titleserv, Inc., 402 F.3d 119 (2d Cir. 2005).

49. See Anna Bernasek and D. T. Mongan, *All You Can Pay* (New York: Nation Books, 2015).

50. David P. Conway, Google Inc., assignee, Dynamic Pricing of Electronic Content. U.S. Patent 8,260,657, filed September 30, 2011, and issued September 4, 2012.

51. Mark Sullivan, "Facebook Patents Technology to Help Lenders Discriminate against Borrowers Based on Social Connections," *Venture Beat*, August 4, 2015, http://venturebeat.com/2015/08/04/facebook-patents-technology-to-help-lenders-discriminate-against-borrowers-based-on-social-connections, accessed November 29, 2015.

52. Andrea Freeman, "Payback: A Structural Solution to the Credit Card Problem," *Arizona Law Review* 55 (Spring 2013): 151–199, at 156.

53. *From Poverty, Opportunity* (Washington, DC: Brookings Institution Metropolitan Poverty Program, 2006), http://www.brookings.edu/research/reports/2006/07/poverty-fellowes, accessed July 7, 2015.

54. Julia Angwin, "The Tiger Mom Tax: Asians Are Nearly Twice as Likely to Get a Higher Price from Princeton Review," *ProPublica*, September 1, 2015, http://www.propublica.org/article/asians-nearly-twice-as-likely-to-get-higher-price-from-princeton-review, accessed November 19, 2015.

55. *Big Data: Seizing Opportunities, Preserving Values* (Washington, DC: Executive Office of the President, 2014), 46, https://www.whitehouse.gov/sites/default/files/docs/big_data_privacy_report_5.1.14_final_print.pdf, accessed July 7, 2015.

56. Louis Phlips, *The Economics of Price Discrimination* (Cambridge, UK: Cambridge University Press, 1983), 18.

57. Joseph Turow, Lauren Feldman, and Kimberly Meltzer, *Open to Exploitation: America's Shoppers Online and Offline* (Philadelphia: University of Pennsylvania Annenberg Public Policy Center, 2005), http://repository.upenn.edu/cgi/viewcontent.cgi?article=1035&context=asc_papers, accessed July 7, 2015.

58. See Wendy Gordon, "Intellectual Property as Price Discrimination: Implications for Contract," *Chicago-Kent Law Review* 73 (1998): 1367–1390, at 1383–1390 (noting that secondary markets are often much better at price discrimination than single monopolistic ones).

59. Kate Abnett, "Will Mass Customization Work for Fashion?" *Business of Fashion*, September 3, 2015, http://www.businessoffashion.com/articles/intelligence/mass-customisation-fashion-nike-converse-burberry, accessed November 29, 2015.

60. List of Crest Toothpaste Products, http://crest.com/en-us/products/toothpaste, accessed November 29, 2015.

61. See Barry Schwartz, *The Paradox of Choice: Why More Is Less* (New York: HarperCollins, 2003); Sheena S. Iyengar and Mark R. Lepper, "When Choice Is Demotivating: Can One Desire Too Much of a Good Thing?" *Journal of Personality and Social Psychology* 79 (2000): 995–1006.

5 The "Buy Now" Lie

1. See Lanham Act, 15 U.S.C. §§ 1051–1141n (2013).

2. Ibid., § 43, 15 U.S.C. § 1125; Federal Trade Commission Act § 5, 15 U.S.C. § 45 (2013).

3. "iTunes Store Terms and Conditions," Apple, Inc., last modified June 30, 2015, http://www.apple.com/legal/internet-services/itunes/us/terms.html#SERVICE, accessed August 21, 2015.

4. "Kindle Store Terms of Use," Amazon Digital Services, Inc., last modified September 6, 2012, http://www.amazon.com/gp/help/customer/display.html?nodeId=201014950, accessed June 14, 2015.

5. "Amazon Music Terms of Use," Amazon Digital Services, Inc., last modified June 11, 2014, http://www.amazon.com/gp/help/customer/display.html?nodeId=201380010, accessed August 21, 2015.

6. Laura Hudson, "For the First Time, You Can Actually Own the Digital Comics You Buy," *Wired*, July 2, 2013, http://www.wired.com/2013/07/drm-free-comics-download-image/, accessed August 21, 2015.

7. Ibid.

8. "Terms and Conditions," Image Comics, https://imagecomics.com/about/terms-and-conditions, accessed August 21, 2015.

9. "Digital Ownership," HeinOnline, http://home.heinonline.org/services/ownership/, accessed August 21, 2015.

10. Michael Kelley, "Random House Says Libraries Own Their Ebooks," *LJ Insider* (blog), *Library Journal*, October 18, 2012, http://lj.libraryjournal.com/2012/10/opinion/random-house-says-libraries-own-their-ebooks-lj-insider, accessed August 21, 2015.

11. Peter Brantley, "Random House Did Not Mean Own, Exactly," *PWxyz* (blog), *Publisher's Weekly*, October 23, 2012, https://web.archive.org/web/20150626112010/http://blogs.publishersweekly.com/blogs/PWxyz/2012/10/23/just-another-word/ (site discontinued), accessed August 21, 2015.

12. "The shop.oreilly.com Ebook Advantage," O'Reilly Media, Inc., n.d., http://shop.oreilly.com/category/ebooks.do, accessed August 21, 2015.

13. "Ebook Usage, Devices, and Formats," O'Reilly Media, Inc., n.d., http://shop.oreilly.com/category/customer-service/ebooks.do, accessed August 21, 2015.

14. In addition, states have adopted their own consumer protection laws prohibiting deceptive trade practices such as false or misleading advertisements. And most

outlaw unfair acts or practices. See, e.g., California Business & Professions Code § 17200.

15. Lanham Act § 43(a). Courts interpreted section 43(a) as creating a sui generis tort of false advertising. Although not all courts were quick to reach that conclusion, they eventually reached something approaching consensus. With the Trademark Law Revision Act of 1988, Congress codified the prevailing judicial reading, dividing section 43(a) into two subsections. The first establishes liability for the infringement of unregistered marks and trade dress. The second creates claims for false advertising and product disparagement. See Trademark Law Revision Act of 1988, Pub. L. No. 100-667, § 132, 102 Stat. 3935, 3946 (1988) (current version at 15 U.S.C. § 1125[a] [2013]).

16. Federal Trade Commission Act § 5(a)(1), 15 U.S.C. § 45(a)(1) (2013).

17. Bruce Schneier, "Do You Know Where Your Data Are?" *Wall Street Journal*, April 28, 2009, http://www.wsj.com/articles/SB123997522418329223, accessed August 21, 2015.

18. Lanham Act §43(a).

19. See Lexmark Int'l, Inc. v. Static Control Components, Inc., 134 S. Ct. 1377 (2014).

20. In one recent example, the court dismissed a false advertising claim brought on behalf of consumers, who alleged that they purchased crab cakes labeled as "Made in the USA" that were actually manufactured with nondomestic crabmeat. See Made in the USA Found. v. Phillips Foods, Inc., 365 F.3d 278 (4th Cir. 2004).

21. See Colligan v. Activities Club of N.Y., Ltd., 442 F.2d 686, 693 (2d Cir. 1971); Coca-Cola Co. v. Procter & Gamble Co., 822 F.2d 28, 31 (6th Cir. 1987) ("Competitors have the greatest interest in stopping misleading advertising, and ... section 43(a) allows those parties with the greatest interest in enforcement, and in many situations with the greatest resources to devote to a lawsuit, to enforce the statute rigorously"); Alpo Petfoods, Inc. v. Ralston Purina Co., 720 F. Supp. 194, 212 (D.D.C. 1989), *aff'd in part, rev'd in part*, 913 F.2d 958 (D.C. Cir. 1990) ("While the Act is not directly available to consumers, it is nevertheless designed to protect consumers, by giving the cause of action to competitors who are prepared to vindicate the injury caused to consumers.").

22. See AT&T Mobility LLC v. Concepcion, 563 U.S. 333, 343 (2011); Am. Exp. Co. v. Italian Colors Rest., 133 S. Ct. 2304, 2309 (2013).

23. See Federal Trade Commission Act § 5(b).

24. Ibid.

25. Johnson & Johnson v. Smithkline Beecham Corp., 960 F.2d 294, 298 (2d Cir. 1992) (plaintiff must show that "a statistically significant part of the commercial

audience holds the false belief allegedly communicated by the challenged advertisement").

26. McNeilab, Inc. v. Am. Home Prods. Corp., 675 F. Supp. 819, 825 (S.D.N.Y. 1987).

27. POM Wonderful, LLC v. FTC, 777 F.3d 478, 490 (D.C. Cir. 2015) ("The Commission ... considers whether 'at least a significant minority of reasonable consumers' would 'likely' interpret the ad to assert the claim") (quoting *In re* Telebrands Corp., 140 F.T.C. 278, 291 [2005]).

28. William H. Morris Co. v. Group W, Inc., 66 F.3d 255, 258 (9th Cir. 1995) (3 percent is not proof that a "significant portion" were deceived); Johnson & Johnson-Merck Consumer Pharms. Co. v. Rhone-Poulenc Rorer Pharms., 19 F.3d 125 (3d Cir. 1994) (survey showing 7.5 percent were deceived or misled was not sufficient).

29. Firestone Tire & Rubber Co. v. FTC, 481 F.2d 246, 249 (6th Cir. 1973) (finding "it hard to overturn the deception findings of the Commission if the ad thus misled 15 percent (or 10 percent) of the buying public"); Novartis Consumer Health, Inc. v. Johnson & Johnson-Merck Consumer Pharms. Co., 290 F.3d 578, 594 (3d Cir. 2002) (survey showing 15 percent of respondents were misled is sufficient to prove a likelihood of deception among a "substantial portion" of the intended audience); Telebrands Corp. v. Media Grp., Inc., 45 U.S.P.Q.2d (BNA) 1342, 1348, 1997 WL 790576, *21–22 (S.D.N.Y. Dec. 23, 1997) (survey showing that 20 percent of the viewers took away the false message is sufficient).

30. James C. Miller, chairman of the FTC, to John D. Dingell, chairman, H.R. Comm. on Energy and Commerce, October 14, 1983, cited in *In re* Cliffdale Assocs., Inc., 103 F.T.C. 110, 175 (1984) ("To be considered reasonable, the interpretation or reaction does not have to be the only one. When a seller's representation conveys more than one meaning to reasonable consumers, one of which is false, the seller is liable for the misleading interpretation.").

31. Charles of the Ritz Distribs. Corp. v. FTC, 143 F.2d 676, 679 (2d Cir. 1944) (quoting Florence Mfg. Co. v. J. C. Dowd & Co., 178 F. 73, 75 [2d Cir. 1910]).

32. James C. Miller to John D. Dingell, October 14, 1983: "A 'material' misrepresentation or practice is one which is likely to affect a consumer's choice of or conduct regarding a product."

33. See James B. Astrachan et al., "An Overview of the Tools Available to the Federal Trade Commission," in *The Law of Advertising*, vol. 5, § 18.01 (Newark, NJ: Matthew Bender, 2015).

34. Aaron Perzanowski and Chris Hoofnagle, "What We Buy When We *Buy Now*," forthcoming *University of Pennsylvania Law Review*, on file with authors.

35. The sample was also roughly in line with the U.S. population in terms of race and geographic distribution.

36. Those terms provided in relevant part:

Upon your download of MediaShop Content and payment of any applicable fees (including applicable taxes), the Content Provider grants you a non-exclusive right to view, use, and display such MediaShop Content an unlimited number of times, solely on the MediaShop or a Reading Application or as otherwise permitted as part of the Service, solely on the number of Supported Devices specified in the MediaShop Store, and solely for your personal, non-commercial use. MediaShop Content is licensed, not sold, to you by the Content Provider. The Content Provider may include additional terms for use within its MediaShop Content. ...

Unless specifically indicated otherwise, you may not sell, rent, lease, distribute, broadcast, sublicense, or otherwise assign any rights to the MediaShop Content or any portion of it to any third party, and you may not remove or modify any proprietary notices or labels on the MediaShop Content. In addition, you may not bypass, modify, defeat, or circumvent security features that protect the MediaShop Content. ...

Termination. Your rights under this Agreement will automatically terminate if you fail to comply with any term of this Agreement. In case of such termination, you must cease all use of the MediaShop Store and the MediaShop Content, and MediaShop may immediately revoke your access to the MediaShop Store and the MediaShop Content without refund of any fees. MediaShop's failure to insist upon or enforce your strict compliance with this Agreement will not constitute a waiver of any of its rights.

37. Yannis Bakos, Florencia Marotta-Wurgler, and David R. Trossen, "Does Anyone Read the Fine Print? Consumer Attention to Standard Form Contracts," New York University Law and Economics Working Papers, Paper 195, New York University School of Law, New York, 2014, 22, http://lsr.nellco.org/cgi/viewcontent.cgi?article=1199&context=nyu_lewp, accessed July 7, 2015 (finding that only one or two out of every thousand retail software shoppers access license agreements).

38. Ben Sheffner of the Motion Picture Association of America has claimed, "If you ask people when you go to a site to buy a movie or a book or a song, I think they pretty much understand that you're not actually buying the copyright. What you are doing is you're purchasing or buying a license which permits you to do certain things." And Catherine Bridge of Disney has stated, "I don't think use of the buy button is a deception, and I think ... that people consuming content are understanding that it's not a physical ownership model." See Department of Commerce Internet Policy Task Force, White Paper on Remixes, First Sale, and Statutory Damages, 56–57n35, January 2016, http://www.uspto.gov/sites/default/files/documents/copyrightwhitepaper.pdf, accessed April 10, 2016.

39. Tim Cushing, "Barnes & Noble Decides That Purchased Ebooks Are Only Yours Until Your Credit Card Expires," *Techdirt*, November 27, 2012, https://www.techdirt.com/articles/20121126/18084721154/barnes-noble-decides-that-purchased-ebooks-are-only-yours-until-your-credit-card-expires.shtml, accessed August 21, 2015.

40. Consider the following language from the Kindle Store's Terms of Use:

Risk of Loss. Risk of loss for Kindle Content transfers when you download or access the Kindle Content.

Termination. Your rights under this Agreement will automatically terminate if you fail to comply with any term of this Agreement. In case of such termination, you must cease all use of the Kindle Store and the Kindle Content, and Amazon may immediately revoke your access to the Kindle Store and the Kindle Content without refund of any fees. Amazon's failure to insist upon or enforce your strict compliance with this Agreement will not constitute a waiver of any of its rights. ("Kindle Store Terms of Use," Amazon Digital Services, Inc.)

41. Not some older devices. And terms are still limited to "Supported Devices." "Supported Device" means a mobile, computer, or other supported electronic device other than a Kindle on which you are authorized to operate a "Reading Application."

42. Kelly Clay, "Amazon Confirms It Makes No Profit on Kindles," *Forbes*, October 12, 2012, http://www.forbes.com/sites/kellyclay/2012/10/12/amazon-confirms-it-makes-no-profit-on-kindles/, accessed August 21, 2015; Steve Kovach, "Amazon Will Lose Millions Selling the Kindle Fire, But That's the Point," *Business Insider*, September 30, 2011, http://www.businessinsider.com/kindle-fire-profit-margins-2011-9, accessed August 21, 2015.

43. Oscar Williams-Grut, "Apple's iPhone: The Most Profitable Product in History," *Independent* (UK), January 29, 2015, http://www.independent.co.uk/news/business/analysis-and-features/apples-iphone-the-most-profitable-product-in-history-10009741.html, accessed August 21, 2015; Daisuke Wakabayashi, "Apple's Market Cap Loses $60 Billion After iPhone Sales Disappoint," *Wall Street Journal*, July 22, 2015, http://www.wsj.com/articles/apple-earnings-boosted-by-iphone-sales-1437510647, accessed August 21, 2015.

44. "iCloud: Family Sharing," Apple, Inc., http://www.apple.com/icloud/family-sharing/, accessed August 25, 2015; see also "Steam Family Sharing," Valve Corporation, http://store.steampowered.com/promotion/familysharing, accessed August 25, 2015.

45. Geoffrey A. Fowler, "Facebook Heir? Time to Choose Who Manages Your Account When You Die," *Wall Street Journal*, February 12, 2015, http://www.wsj.com/articles/facebook-heir-time-to-choose-who-manages-your-account-when-you-die-1423738802, accessed August 25, 2015.

46. See Fiduciary Access to Digital Assets and Digital Accounts Act, 79 Del. Laws 416 (2014), http://www.legis.delaware.gov/LIS/lis147.nsf/vwLegislation/HB%20345/$file/legis.html, accessed August 25, 2015; Uniform Fiduciary Access to Digital Assets Act (Nat'l Conf. of Comm'rs on Unif. State Laws 2014), http://www.uniformlaws.org/shared/docs/Fiduciary%20Access%20to%20Digital%20Assets/2014_UFADAA_Final.pdf, accessed August 25, 2015.

47. Fiduciary Access to Digital Assets and Digital Accounts Act §§ 5004–5005.

48. Ibid., § 5004.

49. That case is even stronger when gender is taken into account. As is often the case with online surveys, women were overrepresented. And since men were

significantly more confused about their resale rights—the same 10 percent for books, but 13.5 percent for music and 25 percent for movies—a more representative gender distribution would have likely shown greater overall deception.

50. Since most readers don't rely on any device other than their eyes or a pair of glasses when reading a hardcover book, the survey asked respondents whether they would prefer a book they can read at the location of their choice. This is an imperfect substitute for the device-of-choice inquiry, but it yielded nearly indistinguishable results.

51. The study reduced the impact of outliers by capping the maximum response for each right at $20.

52. Steven Tweedie, "Apple Made a Small but Significant Change to 'Free' Apps in the App Store," *Business Insider*, November 20, 2014, http://www.businessinsider.com/apple-changes-free-apps-to-get-in-app-store-2014-11, accessed August 25, 2015.

53. See Nathaniel Good, Jens Grossklags, David Thaw, Aaron Perzanowski, Deirdre Mulligan, and Joseph Konstan, "User Choices and Regret: Understanding Users' Decision Process about Consensually Acquired Spyware," *I/S: A Journal of Law and Policy* 2, no. 2 (2006): 283–344; Joel R. Reidenberg, Travis Breaux, Lorrie Faith Cranor, Brian French, Amanda Grannis et. al., "Disagreeable Privacy Policies: Mismatches Between Meaning and Users' Understanding," *Berkeley Technology Law Journal* 30 (2015): 39–87, at 48; Lorrie Faith Cranor, "Necessary but Not Sufficient: Standardized Mechanisms for Privacy Notice and Choice," *Journal of Telecommunications & High Technology Law* 10, (2012): 273–307, at 293; Marie C. Pollio, "The Inadequacy of HIPAA's Privacy Rule: The Plain Language Notice of Privacy Practices and Patient Understanding," N.Y.U. Annual Survey of American Law 60, (2004): 579–620, at 615; Katy K. Liu, "Fair and Accurate Credit Transactions Act Regulations: Disclosure, Opt-Out Rights, Medical Information Usage, and Consumer Information Disposal," *I/S: A Journal of Law and Policy* 2, (2006): 715–735, at 720.

6 The Promise and Perils of Digital Libraries

1. Others contend the first major public library was the Boston Public Library, founded in 1852 through a large donation from Joshua Bates, because it was the first to make all of its books "free to all" Boston citizens. See John Palfrey, *BiblioTech: Why Libraries Matter More Than Ever in the Age of Google* (New York: Basic Books, 2015), 1.

2. See, e.g., N.Y. C.P.L.R. § 4509 (Consol. 2015) ("Library records, which contain names or other personally identifying details regarding the users of public, free association, school, college and university libraries and library systems of this state, including but not limited to records related to the circulation of library materials, computer database searches, interlibrary loan transactions, reference queries,

requests for photocopies of library materials, title reserve requests, or the use of audio-visual materials, films or records, shall be confidential and shall not be disclosed except that such records may be disclosed to the extent necessary for the proper operation of such library and shall be disclosed upon request or consent of the user or pursuant to subpoena, court order or where otherwise required by statute."); "Code of Ethics of the American Library Association," last amended January 22, 2008, http://www.ala.org/advocacy/proethics/codeofethics/codeethics, accessed August 6, 2015; "Privacy: An Interpretation of the Library Bill of Rights," American Library Association, adopted June 19, 2002, http://www.ala.org/Template.cfm?Section=interpretations&Template=/ContentManagement/ContentDisplay.cfm&ContentID=132904, accessed August 6, 2015. See also Adam L. Penenberg, "Don't Mess with Librarians," *Wired*, September 15, 2004, http://www.wired.com/2004/09/dont-mess-with-librarians/, accessed August 6, 2015.

3. Neil Richards, *Intellectual Privacy: Rethinking Civil Liberties in the Digital Age* (New York: Oxford University Press, 2015). See also "DRAFT—Library Privacy Guidelines for E-book Lending and Digital Content Vendors," American Library Association, last modified June 16, 2015, http://connect.ala.org/node/241070#sthash.OUjmCDa7.dpuf, accessed August 6, 2015: "Protecting user privacy and confidentiality has long been an integral part of the intellectual freedom mission of libraries."

4. Boston Public Library's McKim Building bears the inscription "FREE-TO-ALL" above its main entrance—an expression of one of the longstanding values of the BPL system. For a photograph of the inscription, see http://maps.bpl.org/sites/default/files/images/about_ph_6.jpg?1361904290, accessed August 6, 2015.

5. David O'Brien, Urs Gasser, and John G. Palfrey Jr., "E-Books in Libraries: A Briefing Document Developed in Preparation for a Workshop on E-Lending in Libraries," Berkman Center Research Publication No. 2012-15, Berkman Center for Internet & Society, Harvard University, Cambridge, MA, 2012, https://cyber.law.harvard.edu/publications/2012/ebooks_in_libraries, accessed August 6, 2015.

6. Ted Striphas, *The Late Age of Print: Everyday Book Culture from Consumerism to Control* (New York: Columbia University Press, 2009), 35.

7. K. T. Bradford, "Paper Rules: Why Borrowing an E-book from Your Library Is So Difficult," *Digital Trends*, June 15, 2013, http://www.digitaltrends.com/mobile/e-book-library-lending-broken-difficult/, accessed August 6, 2015; Simon & Schuster CEO Carolyn Reidy revealed some of the thought process behind the publisher's new library ebook pilot program: "If you could get every book you wanted free, why would you ever buy another one?"; Jay Greene, "Penguin Halts Libraries' E-book Lending, Citing Security Fears," *CNET*, November 22, 2011, updated November 23, 2011, http://www.cnet.com/news/penguin-halts-libraries-e-book-lending-citing-security-fears/, accessed August 6, 2015; "New Zealand Publisher Fears Libraries," *Annoyed Librarian* (blog), *Library Journal*, May 18, 2009, http://lj.libraryjournal.com/blogs/annoyedlibrarian/2009/05/18/new-zealand-publisher-fears-libraries/,

accessed August 6, 2015; Ken Chad, "'Frictionless' Ebook Lending from Public Libraries," *Shelf Free* (blog), July 26, 2013, http://shelffree.org.uk/2013/07/26/frictionless-ebook-lending-from-public-libraries/, accessed August 6, 2015; David R. O'Brien, Urs Gasser, and John Palfrey, *E-Books in Libraries* (Cambridge, MA: The Berkman Center for Internet & Society at Harvard University, 2012), PDF briefing document, 22, http://cyber.law.harvard.edu/sites/cyber.law.harvard.edu/files/E-Books%20in%20Libraries%20(O'Brien,%20Gasser,%20Palfrey)-1.pdf, accessed August 6, 2015.

8. Kathy Rosa, *Research and Statistics on Libraries and Librarianship in 2013* (Chicago: American Library Association, 2014), PDF report, http://www.ala.org/research/sites/ala.org.research/files/content/librarystats/LBTA-research2014.pdf, accessed August 6, 2015.

9. "E-books, U.S. Public Libraries 2012," American Library Association, http://www.ala.org/research/plftas/2011_2012/ebooksmap, accessed August 6, 2015.

10. Eckhard Kummrow, "PwC Prognosis: 11% Sales Volume of eBooks in Germany by 2018; 52% in the USA," *Librarian in Residence* (blog), November 25, 2014, http://blog.goethe.de/librarian/index.php?archives/540-English.html, accessed August 6, 2015. But see Frank Catalano, "Paper Is Back: Why 'Real' Books Are on the Rebound," *GeekWire*, January 18, 2015, http://www.geekwire.com/2015/paper-back-real-books-rebound/, accessed March 15, 2016.

11. "Children's eBook Sales Surge in 2014 in Public Libraries and Schools through Overdrive," *OverDrive Blogs* (blog), January 13, 2015, http://blogs.overdrive.com/library/2015/01/13/childrens-ebook-sales-surge-in-2014-in-public-libraries-and-schools-through-overdrive/, accessed August 6, 2015.

12. Ibid.

13. Alexandra Alter, "The Plot Twist: E-Book Sales Slip, and Print Is Far from Dead," *New York Times*, September 22, 2015, http://www.nytimes.com/2015/09/23/business/media/the-plot-twist-e-book-sales-slip-and-print-is-far-from-dead.html, accessed November 20, 2015.

14. Note that ReDigi and others have tried to address this concern by migrating copies from one device to another instead of copying them in a classic sense. Because only one version of the file exists at any given moment, their techniques are far more analogous to moving a copy or restoring/repairing a copy than reproducing one.

15. See R. Anthony Reese, "The First Sale Doctrine in the Era of Digital Networks," University of Texas Law, Public Law Research Paper No. 57; and University of Texas Law, Law and Economics Research Paper No. 004, Austin, TX, 2003, http://papers.ssrn.com/sol3/papers.cfm?abstract_id=463620, accessed August 6, 2015.

16. *DMCA Section 104 Report* (Washington, DC: U.S. Copyright Office, 2001), PDF report, 88, http://www.copyright.gov/reports/studies/dmca/sec-104-report-vol-1.pdf, accessed August 6, 2015.

17. One study estimated that from 1994 to 2005, library use of license agreement rose 600 percent. See Sharon Farb, "Libraries, Licensing, and the Challenge of Stewardship," *First Monday* 11, no. 7 (2006), doi: 10.5210/fm.v11i7.1364.

18. Amy Calhoun and James English, "Library Simplified" (presentation delivered at ALA 2015, San Francisco, CA, June 27, 2015), http://www.slideshare.net/jamesenglish/library-simplified-ala, accessed November 20, 2015.

19. Ava Seave, "Are Digital Libraries a 'Winner-Takes-All' Market? OverDrive Hopes So," *Forbes*, November 18, 2013, http://www.forbes.com/sites/avaseave/2013/11/18/are-digital-libraries-a-winner-takes-all-market-overdrive-hopes-so/, accessed August 6, 2015.

20. While there hasn't been a documented case of a library patron having their ebook access cut off, similar situations have arisen for Amazon ebook purchasers. See, e.g., Mark King, "Amazon Wipes Customer's Kindle and Deletes Account with No Explanation," *Guardian* (UK), October 22, 2012, http://www.theguardian.com/money/2012/oct/22/amazon-wipes-customers-kindle-deletes-account, accessed August 6, 2015.

21. Josh Hadro, "HarperCollins Puts 26 Loan Cap on Ebook Circulations," *Library Journal*, February 25, 2011, http://lj.libraryjournal.com/2011/02/technology/ebooks/harpercollins-puts-26-loan-cap-on-ebook-circulations/, accessed August 6, 2015. In response, one library then posted video of one of its analog HC copies after twenty-six lends to show that it was in near-perfect condition. Cory Doctorow, "How a HarperCollins Library Book Looks after 26 Checkouts (Pretty Good!)," *Boing Boing*, March 3, 2011, http://boingboing.net/2011/03/03/how-a-harpercollins.html, accessed August 6, 2015.

22. Michael Kelley, "One Year Later, HarperCollins Sticking to 26-Loan Cap, and Some Librarians Rethink Opposition," *Digital Shift*, February 17, 2012, http://www.thedigitalshift.com/2012/02/ebooks/one-year-later-harpercollins-sticking-to-26-loan-cap-and-some-librarians-rethink-opposition/, accessed August 6, 2015.

23. Kevin Smith, Planning for Musical Obsolescence, Scholarly Communications @ Duke (blog), July 28, 2014, http://blogs.library.duke.edu/scholcomm/2014/07/28/planning-for-musical-obsolescence/, accessed August 6, 2015. See also "Sound Recording Collecting in Crisis: Home," University of Washington Libraries, last modified February 24, 2016, http://guides.lib.uw.edu/research/imls2014, accessed March 12, 2016.

24. Michael Kozlowski, "Penguin Random House Announces New e-Book Terms for Libraries," *Good E-Reader* (blog), December 6, 2015, http://goodereader.com/blog/e-book-news/penguin-random-house-announces-new-e-book-terms-for-libraries, accessed March 15, 2016.

25. Jennifer M. Urban, "How Fair Use Can Help Solve the Orphan Works Problem," *Berkeley Technology Law Journal* 27 (2012): 1379–1429, http://papers.ssrn.com/sol3/papers.cfm?abstract_id=2089526, accessed August 6, 2015.

26. Amy Kirchhoff, "eBooks: The Preservation Challenge," *Against the Grain* 23, no. 4 (2011): 32, http://docs.lib.purdue.edu/cgi/viewcontent.cgi?article=5935&context=atg, accessed August 6, 2015.

27. See Letter from Mary Koelbel Engle, Associate Director, Division of Advertising Practices, Bureau of Consumer Protection, Federal Trade Commission to M. Sean Royall, Esq., "Re: Wal-Mart Stores, Inc., FTC File No. 092-3003," June 23, 2010, https://www.ftc.gov/sites/default/files/documents/closing_letters/wal-mart-stores-inc./100623walmartletter.pdf, accessed August 6, 2015.

28. David Gary, "Saving the Scream Queens," *Atlantic*, August 21, 2015, http://www.theatlantic.com/entertainment/archive/2015/08/saving-the-scream-queens/401141/, accessed March 15, 2016.

29. Ibid.

30. See "Banned Books That Shaped America," Banned Books Week, http://www.bannedbooksweek.org/censorship/bannedbooksthatshapedamerica, accessed August 6, 2015.

31. "Mein Kampf to Be Republished in Germany in Early 2016," *Telegraph* (UK), February 25, 2015, http://www.telegraph.co.uk/news/worldnews/europe/germany/11433843/Mein-Kampf-to-be-republished-in-Germany-in-early-2016.html, accessed August 6, 2015.

32. 227 F.3d 1110 (9th Cir. 2000).

33. The U.S. Copyright Act does allow copyright owners to impound or destroy infringing copies that are in possession of the infringer before they are sold, but not from third parties post-sale. See 17 U.S.C. § 503 (2013).

34. Bill Chappell, "Boy Says He Didn't Go to Heaven; Publisher Says It Will Pull Book," *The Two-Way, NPR*, January 15, 2015, http://www.npr.org/sections/the two-way/2015/01/15/377589757/boy-says-he-didn-t-go-to-heaven-publisher-says-it-will-pull-book, accessed March 15, 2016.

35. Ron Charles, "'Boy Who Came Back from Heaven' Actually Didn't; Books Recalled," *Washington Post*, January 16, 2016, https://www.washingtonpost.com/news/arts-and-entertainment/wp/2015/01/15/boy-who-came-back-from-heaven-going-back-to-publisher/, accessed March 15, 2016.

36. See Marc Blitz, "Constitutional Safeguards for Silent Experiments in Living: Libraries, the Right to Read, and a First Amendment Theory for an Unaccompanied Right to Receive Information," *University of Missouri-Kansas City Law Review* 74 (2006): 799–882.

37. For a detailed argument on the dangers of tracking reader usage, see Privacy Authors and Publishers' Objection to Proposed Settlement, Author's Guild v. Google, Inc., no. 1:05-CV-08136-DC (S.D.N.Y. Sept. 8, 2009), ECF No. 281, https://www.eff.org/files/filenode/authorsguild_v_google/file_stamped_brf.pdf, accessed April 10, 2016; Reader Privacy Act, 2011 Cal. Adv. Legis. Serv. 424 (LexisNexis), http://www.leginfo.ca.gov/pub/11-12/bill/sen/sb_0601-0650/sb_602_bill_20111002_chaptered.html, accessed August 6, 2015.

38. See Corynne McSherry, "Adobe Spyware Reveals (Again) the Price of DRM: Your Privacy and Security," *Deeplinks* (blog), Electronic Frontier Foundation, October 7, 2014, https://www.eff.org/deeplinks/2014/10/adobe-spyware-reveals-again-price-drm-your-privacy-and-security, accessed August 6, 2015.

39. See *Consensus Framework to Support Patron Privacy in Digital Library and Information Systems*, prepared by the National Information Standards Organization (Baltimore, 2015), http://www.niso.org/topics/tl/patron_privacy/, accessed April 10, 2016.

40. See Riley v. California, 134 S. Ct. 2473 (2014); United States v. Jones, 132 S. Ct. 945 (2012); Katz v. United States, 389 U.S. 347 (1967).

41. See, e.g., Tattered Cover, Inc. v. City of Thornton, 44 P.3d 1044 (Colo. 2002) (en banc), http://caselaw.findlaw.com/co-supreme-court/1340412.html, accessed April 10, 2016.

42. Smith v. Maryland, 442 U.S. 735 (1979); United States v. Miller, 425 U.S. 435 (1976).

43. United States v. Rumley, 345 U.S. 41, 57 (1953) (J. Douglas, concurring).

44. See Exec. Sess. of the S. Perm. Subcomm. on Investigations of the Comm. on Gov't Operations, 83d Cong. 964 (1953) (testimony of Jerre G. Mangione).

45. Ibid. at 1697 (testimony of Mary Colombo Palmiero).

46. Lamont v. Postmaster Gen., 381 U.S. 301, 307 (1965).

47. David Streitfeld, "Kramerbooks Vows to Stand Firm," *Washington Post*, May 29, 1998, http://www.washingtonpost.com/wp-srv/politics/special/clinton/stories/kramer052998.htm, accessed August 6, 2015. See also *In re* Grand Jury Subpoena to Kramerbooks & Afterwords, Inc., 26 Media L. Rep. (BNA) 1599, 1601 (D.D.C. 1998) (finding that as a result of a grand jury subpoena for a patron's book purchases, "many customers have informed Kramerbooks personnel that they will no longer

shop at the bookstore because they believed Kramerbooks to have turned documents over ... that reveal a patron's choice of books").

48. "Bookstore Confidentiality Preserved in Lewinsky Case: A 'Chilling' Standoff Comes to an End," *PR Newswire*, June 23, 1998, http://www.prnewswire.com/news-releases/bookstore-confidentiality-preserved-in-lewinsky-case-a-chilling-standoff-comes-to-an-end-78064387.html, accessed August 6, 2015; David Streitfeld, "Starr Will Get Bookstore Records," *Washington Post*, June 23, 1998, http://www.washingtonpost.com/wp-srv/politics/special/clinton/stories/starr062398.htm, accessed August 6, 2015.

49. *In re* Grand Jury Subpoena to Amazon.com, 246 F.R.D. 570, 573 (W.D. Wis. 2007).

50. Dawinder S. Sidhu, "The Chilling Effect of Government Surveillance Programs on the Use of the Internet by Muslim-Americans," *University of Maryland Law Journal of Race, Religion, Gender and Class* 7, no. 2 (2007): 375–393, at 391, http://digitalcommons.law.umaryland.edu/cgi/viewcontent.cgi?article=1134&context=rrgc, accessed September 3, 2015. See also ACLU v. Gonzales, 478 F. Supp. 2d 775, 805–806 (E.D. Pa. 2007) (finding that "many people wish to browse and access material privately and anonymously, especially if it is sexually explicit," that "as a result of this desire to remain anonymous, many users who are not willing to access information non-anonymously will be deterred from accessing the desired information," and that "web site owners such as the plaintiffs will be deprived of the ability to provide this information to those users"); United States v. Curtin, 489 F.3d 935, 959 (9th Cir. 2007) (Kleinfeld, J., concurring) (noting that "in the 1950s, people with leftist books sometimes shelved them spine to the wall, out of fear that visitors would see and report them").

51. Between 2001 and 2005, libraries were contacted by law enforcement seeking information on patrons at least two hundred times. Eric Lichtblau, "F.B.I., Using Patriot Act, Demands Library's Records," *New York Times*, August 26, 2005, http://www.nytimes.com/2005/08/26/politics/fbi-using-patriot-act-demands-librarys-records.html, accessed September 3, 2015.

52. City of Los Angeles v. Patel, 135 S. Ct. 2443 (2015), http://www.supremecourt.gov/opinions/14pdf/13-1175_2qe4.pdf, accessed September 3, 2015 (finding that hotels have a right to object to government searches for information about their guests).

53. Quoted in Steven R. Harris, "Mortgaging Our Future on Ownership, or, the Pleasures of Renting," *Against the Grain* 23, no. 4 (2011): 28, http://docs.lib.purdue.edu/cgi/viewcontent.cgi?article=5934&context=atg, accessed September 4, 2015.

54. For more information on these endeavors, see Open Library's website, https://openlibrary.org/, accessed September 4, 2015, and the Digital Public Library of America's website, http://dp.la/, accessed September 4, 2015.

55. Additional information on NYPL Labs can be found at http://www.nypl.org/collections/labs, accessed September 4, 2015.

56. HathiTrust, https://www.hathitrust.org/, accessed September 4, 2015.

57. See 17 U.S.C. § 107.

58. Project Cicero, http://www.projectcicero.org/, accessed September 4, 2015.

7 DRM and the Secret War inside Your Devices

1. Kyle Wiens, "e-Book Legal Restrictions Are Screwing Over Blind People," *Wired*, December 15, 2014, http://www.wired.com/2014/12/e-books-for-the-blind-should-be-legal/, accessed September 5, 2015.

2. Eric Bangeman, "DirecTV DVR Clampdown: A Sober Reminder of DRM Suckitude," *Ars Technica* (blog), March 20, 2008, http://arstechnica.com/uncategorized/2008/03/directv-dvr-clampdown-a-sober-reminder-of-drm-suckitude/, accessed September 5, 2015.

3. "Lending for Kindle," Kindle Direct Publishing, https://kdp.amazon.com/help?topicId=A2P1X97KAW8GZE, accessed September 5, 2015.

4. Cory Doctorow, *Information Doesn't Want to Be Free: Laws for the Internet Age* (San Francisco: McSweeney's, 2014), under sec. 1.1, "Anti-Circumvention Explained": "Ever tried to fast-forward through the anti-piracy warning at the start of a DVD and gotten an action not allowed message? That's a digital lock." Ibid. "With digital locks, you can sell them only the right to look at the book after 6 p.m., while physically located in North America and not in a commercial establishment. If they want the 'read on the subway' rights, those can be sold separately."

5. See also Laura Northrup, "Here's Why Digital Rights Management Is Stupid and Anti-Consumer," *Consumerist* (blog), November 26, 2012, http://consumerist.com/2012/11/26/heres-why-digital-rights-management-is-stupid-and-anti-consumer/, accessed September 5, 2015.

6. "Smart Cow Problem," *Wikipedia*, last modified June 3, 2014, https://en.wikipedia.org/wiki/Smart_cow_problem, accessed September 5, 2015.

7. *Home Recording of Copyrighted Works: Hearings on H.R. 4783, H.R. 4794, H.R. 4808, H.R. 5250, H.R. 5488, and H.R. 5705 Before the Subcomm. on Courts, Civil Liberties, and the Admin. of Justice of the H. Comm. on the Judiciary*, 97th Cong. (1982) (testimony of Jack Valenti, president, Motion Picture Association of America, Inc.), http://cryptome.org/hrcw-hear.htm, accessed September 5, 2015.

8. See DVD Copy Control Ass'n, Inc. v. Kaleidescape, Inc., 97 Cal. Rptr. 3d 856 (Cal. Ct. App. 6th Dist. 2009).

9. Sega also claimed that Accolade infringed its trademark because the TMSS code prompted the Genesis to display the Sega logo on screen. The court rejected this theory as well.

10. The statute also requires that copyright holders receive royalty payments for DAT sales.

11. Universal City Studios v. Corley, 273 F.3d 429, 439 (2d Cir. 2001).

12. The right to back up consumer-owned DVDs was also asserted in another important section 1201 case, *321 Studios v. MGM Studios, Inc.*, 307 F. Supp. 2d 1085 (N.D. Cal. 2004). It similarly fell on deaf ears.

13. See http://kotaku.com/the-anti-piracy-tech-thats-tearing-video-game-hackers-a-1759518600, accessed April 10, 2016.

14. Mike Mansick, "Ubisoft's Annoying New DRM Cracked within Hours of Release," *Techdirt*, March 4, 2010, https://www.techdirt.com/articles/20100304/1302148421.shtml, accessed September 5, 2015.

15. See Tim Anderson, "How Apple Is Changing DRM," *Guardian* (UK), May 15, 2008, http://www.theguardian.com/technology/2008/may/15/drm.apple, accessed September 5, 2015.

16. Timothy Geigner, "The Full Counter-Argument to Game Studios Claiming a Need for DRM: *The Witcher 3*," *Techdirt*, August 31, 2015, https://www.techdirt.com/articles/20150827/05171032075/full-counter-argument-to-game-studios-claiming-need-drm-witcher-3.shtml, accessed November 20, 2015.

17. Steve Jobs, "Thoughts on Music," Apple, Inc., February 6, 2007, https://web.archive.org/web/20070207234839/http://www.apple.com/hotnews/thoughtsonmusic/, accessed September 5, 2015.

18. Jeff Elder, "Former iTunes Engineer Tells Court He Worked to Block Competitors," *Digits* (blog), *Wall Street Journal*, December 12, 2014, http://blogs.wsj.com/digits/2014/12/12/former-itunes-engineer-tells-court-he-worked-to-block-competitors/, accessed September 5, 2015. As a matter of disclosure, one of the authors of this book was a paid consultant for the plaintiffs in the antitrust case against Apple.

19. Doctorow, *Information Doesn't Want To Be Free*, 593.

20. This results in what Chris Anderson called "the long tail," where the low costs of distribution allow for works to remain in circulation, even if their audience is small in absolute terms. See Chris Anderson, *The Long Tail* (New York: Hatchette Books, 2006).

21. Eric von Hippel, *Democratizing Innovation* (Cambridge, MA: MIT Press, 2005), http://web.mit.edu/evhippel/www/democ1.htm, accessed September 5, 2015.

22. Ibid.

23. Kevin Poulsen, "Hackers Sued for Tinkering with Xbox Games," *SecurityFocus*, February 9, 2005, http://www.securityfocus.com/news/10466, accessed September 5, 2015.

24. Davidson & Assoc. v. Jung, 422 F.3d 630 (8th Cir. 2005). One of this book's authors represented the developers of bnetd in this case.

25. MDY Indus., LLC v. Blizzard Entm't, Inc., 629 F.3d 928 (9th Cir. 2010).

26. For a detailed analysis of the Sony rootkit fiasco, see Deirdre Mulligan and Aaron Perzanowski, "The Magnificence of the Disaster: Reconstructing the Sony BMG Rootkit Incident," *Berkeley Technology Law Journal* 22, no. 3 (Summer 2007): 1157–1232.

27. Doctorow, *Information Doesn't Want To Be Free*, 779.

28. Chamberlain Grp., Inc. v. Skylink Techs., Inc., 292 F. Supp. 2d 1023, 1039 (N.D. Ill. 2003).

29. Chamberlain Grp., Inc. v. Skylink Techs., Inc., 381 F.3d 1178, 1202 (Fed. Cir. 2004).

30. Lexmark Int'l, Inc. v. Static Control Components, Inc., 387 F.3d 522 (6th Cir. 2004).

8 The Internet of Things You Don't Own

1. Riley v. California, 134 S. Ct. 2474, 2489–90 (2014).

2. Cory Doctorow, "How Laws Restricting Tech Actually Expose Us to Greater Harm," *Wired*, December 24, 2014, http://www.wired.com/2014/12/government-computer-security/, accessed September 7, 2015.

3. Tim Cushing, "DRM; Or How to Make 30,000-Hour LED Bulbs 'Last' Only One Month," *Techdirt*, March 18, 2015, https://www.techdirt.com/articles/20150317/08091030343/drm-how-to-make-30000-hour-led-bulbs.shtml, accessed September 7, 2015.

4. Aaron Smith, "New Ford Car Automatically Obeys Speed Limits," *CNN Money*, March 25, 2015, http://money.cnn.com/2015/03/25/technology/ford-speed-limit/, accessed September 7, 2015.

5. Stuart Dredge, "White House Drone Crash Fallout Shows Who Really Owns Your Drones, Says EFF," *Guardian* (UK), February 3, 2015, http://www.theguardian.com/technology/2015/feb/03/white-house-drone-crash-eff, accessed September 7, 2015.

6. Tim Cushing, "DRM, Or How to Turn Your Cat's Litter Box into an Inkjet Printer," *Techdirt*, January 8, 2015, https://www.techdirt.com/articles/20150102/09574429580/drm-how-to-turn-your-cats-litter-box-into-inkjet-printer.shtml, accessed September 7, 2015.

7. Kashmir Hill, "Samsung Wants You to Put a Motion Tracker under a Loved One's Mattress—What Could Go Wrong?" *Fusion*, September 4, 2015, http://fusion.net/story/193514/samsung-sleepsense-loved-ones-mattress/, accessed November 20, 2015.

8. Rob Price, "The Smart-Home Device That Google Is Deliberately Disabling Was Sold with a 'Lifetime Subscription,'" *Business Insider*, April 5, 2016, http://www.businessinsider.com/revolv-smart-home-hubs-lifetime-subscription-bricked-nest-google-alphabet-internet-of-things-2016-4, accessed April 10, 2016.

9. Arlo Gilbert, "The Time That Tony Fadell Sold Me a Container of Hummus," *Medium*, April 3, 2016, https://medium.com/@arlogilbert/the-time-that-tony-fadell-sold-me-a-container-of-hummus-cb0941c762c1#.nhl96qogu, accessed April 10, 2016.

10. "Apple Reinvents the Phone with iPhone," Apple Press Info, Apple Inc., January 9, 2007, http://www.apple.com/pr/library/2007/01/09Apple-Reinvents-the-Phone-with-iPhone.html, accessed September 7, 2015.

11. Ibid.

12. Hush-A-Phone Corp. v. United States, 238 F.2d 266 (D.C. Cir. 1956).

13. See, e.g., Andrew "bunnie" Huang, *Hacking the Xbox: An Introduction to Reverse Engineering* (San Francisco: No Starch Press, 2003); MythTV, https://www.mythtv.org/, accessed September 7, 2015.

14. See "First Jailbreaks by Device and iOS Version," in "iOS Jailbreaking," *Wikipedia*, last modified September 6, 2015, https://en.wikipedia.org/wiki/IOS_jailbreaking, accessed September 7, 2015.

15. Responsive Comment of Apple Inc. in Opposition to Proposed Exemption 5A and 11A (Class #1), *In re* Exemption to Prohibition on Circumvention of Copyright Protection Systems for Access Control Technologies, No. RM 2008-8 (U.S. Copyright Office, February 2, 2009), https://www.eff.org/files/filenode/dmca_2009/apple-inc-31.pdf, accessed September 7, 2015; Response of Apple Inc. to Questions Submitted by the Copyright Office Concerning Exemptions 5A and 11A (Class #1), *In re* Exemption to Prohibition on Circumvention of Copyright Protection Systems for Access Control Technologies, No. RM 2008-8 (U.S. Copyright Office, July 13, 2009), https://www.eff.org/files/filenode/dmca_2009/apples-response-to-copyright-office-questions-of-6-23-09.pdf, accessed September 7, 2015.

16. Adi Robertson, "As of Today, Americans Can Legally Unlock Their Phones Again," *Verge*, August 1, 2014, http://www.theverge.com/2014/8/1/5959915/president-barack-obama-signing-phone-unlocking-bill, accessed September 7, 2015.

17. "Discovery of Gene" in "History of Roundup Ready Crops," *SourceWatch*, last modified August 19, 2012, http://www.sourcewatch.org/index.php/History_of_Roundup_Ready_Crops#Discovery_of_Gene, accessed September 7, 2015.

18. Bowman v. Monsanto Co., 133 S. Ct. 1761 (2013), https://scholar.google.com/scholar_case?case=14668330492460109241&hl=en&as_sdt=6&as_vis=1&oi=scholarr, accessed September 7, 2015. See also Monsanto Canada Inc. v. Schmeiser, [2004] S.C.R. 902 (Can.), http://scc-csc.lexum.com/scc-csc/scc-csc/en/item/2147/index.do, accessed March 15, 2016. We'll discuss the Monsanto patents and their associated lawsuits later in chapter 9.

19. Deere & Company, *Operator Manual: 9120, 9220, 9320, 9420, and 9520 Tractors* (2003), http://manuals.deere.com/omview/OMAR183678_19/, accessed September 7, 2015.

20. Deere & Company, "Engine Control Unit Service Codes-(ECU)," in *Operator Manual: 9120, 9220, 9320, 9420, and 9520 Tractors* (2003), http://manuals.deere.com/omview/OMAR183678_19/RW24911_0000099_19_05NOV01_1.htm, accessed September 7, 2015.

21. See, e.g., Kyle Wiens, "New High-Tech Farm Equipment Is a Nightmare for Farmers," *Wired*, February 5, 2015, http://www.wired.com/2015/02/new-high-tech-farm-equipment-nightmare-farmers/, accessed September 7, 2015.

22. Long Comment Regarding a Proposed Exemption under 17 U.S.C. 1201 (Proposed Class # 21) at 6, Section 1201 Exemptions to Prohibition against Circumvention of Technological Measures Protecting Copyrighted Works: Second Round of Comments, http://copyright.gov/1201/2015/comments-032715/class%2021/John_Deere_Class21_1201_2014.pdf, accessed September 7, 2015.

23. Aro Mfg. Co. v. Convertible Top Replacement Co., 365 U.S. 336 (1961).

24. Isaac Newton to Robert Hooke, February 15, 1675, in *The Correspondence of Isaac Newton, Vol. 1 (1661–1675)*, ed. H. W. Turnbull (London: Cambridge University Press, 1960), 416.

25. Eric von Hippel, *Democratizing Innovation* (Cambridge, MA: MIT Press, 2005), 2.

26. See Ethan Zuckerman, "Eric von Hippel and 2.9 Million British Innovators," ... *My Heart's in Accra* (blog), September 14, 2010, http://www.ethanzuckerman.com/blog/2010/09/14/eric-von-hippel-and-2-9-million-british-innovators/, accessed September 7, 2015.

27. Wiens, "New High-Tech Farm Equipment Is a Nightmare for Farmers."

28. Comments of General Motors LLC (Proposed Class #21) at 10, Exemption to Prohibition on Circumvention of Copyright Protection Systems for Access Control Technologies, No. 2014-07 (U.S. Copyright Office, March 27, 2015), http://copyright.gov/1201/2015/comments-032715/class%2021/General_Motors_Class21_1201_2014.pdf, accessed November 20, 2015.

29. Michael Corkery and Jessica Silver-Greenberg, "Miss a Payment? Good Luck Moving That Car," *New York Times*, September 24, 2014, http://dealbook.nytimes.com/2014/09/24/miss-a-payment-good-luck-moving-that-car/, accessed September 7, 2015.

30. "mbrace," Mercedes-Benz USA, LLC, https://www.mbusa.com/mercedes/mbrace, accessed September 7, 2015.

31. "Mercedes-Benz mbrace Terms of Service," Mercedes-Benz USA, LLC, last modified September 8, 2013, http://www.mbusa.com/vcm/MB/DigitalAssets/pdfmb/mbraceservicebrochures/mbrace_Terms_of_Service_9-8-13.pdf, accessed September 7, 2015.

32. See "Massachusetts Right to Repair," http://massrighttorepair.com/, accessed September 7, 2015; Jason Torchinsky, "Carmakers Want to Use Copyright Law to Make Working on Your Car Illegal," *Jalopnik* (blog), April 21, 2015, http://jalopnik.com/carmakers-want-to-make-working-on-your-car-illegal-beca-1699132210, accessed September 7, 2015.

33. Torchinsky, "Carmakers Want to Use Copyright Law to Make Working on Your Car Illegal."

34. Alisa Priddle, "Ford Recalls 433,000 Cars Because Engines Won't Shut Off," *USA Today*, July 2, 2015, http://www.usatoday.com/story/money/cars/2015/07/02/ford-recall/29620683/, accessed September 7, 2015.

35. Andy Greenberg, "After Jeep Hack, Chrysler Recalls 1.4M Vehicles for Bug Fix," *Wired*, July 24, 2015, http://www.wired.com/2015/07/jeep-hack-chrysler-recalls-1-4m-vehicles-bug-fix/, accessed September 7, 2015.

36. Bill Chappell, "11 Million Cars Worldwide Have Emissions 'Defeat Device,' Volkswagen Says," *The Two-Way*, *NPR*, September 22, 2015, http://www.npr.org/sections/thetwo-way/2015/09/22/442457697/11-million-cars-worldwide-have-emissions-problem-volkswagen-says, accessed November 21, 2015.

37. Charles Lane, "Emissions Scandal Is Hurting VW Owners Trying to Resell," *NPR*, October 27, 2015, http://www.npr.org/2015/10/26/450238773/emissions-scandal-is-hurting-vw-owners-trying-to-resell, accessed November 21, 2015.

38. Sean O'Kane, "Automakers Just Lost the Battle to Stop You from Hacking Your Car," *Verge*, October 27, 2015, http://www.theverge.com/2015/10/27/9622150/dmca-exemption-accessing-car-software, accessed November 21, 2015.

39. Daniel Cooper, "Ford's New Car Will Force You to Obey the Speed Limit," *Engadget*, March 24, 2015, http://www.engadget.com/2015/03/24/ford-smax-speed-limit/, accessed September 7, 2015.

40. Jim Edwards, "Ford Exec: 'We Know Everyone Who Breaks the Law' Thanks to Our GPS in Your Car," *Business Insider*, January 8, 2014, http://www.businessinsider.com/ford-exec-gps-2014-1#ixzz3ksVKHeH4, accessed September 7, 2015.

41. Tim Cushing, "Ferrari 'DRM:' Don't Screw with Our Logos and We'll Let You Know If It's OK to Sell Your Car," *Techdirt*, September 3, 2014, https://www.techdirt.com/articles/20140902/11491828395/ferrari-drm-dont-screw-with-our-logos-well-let-you-know-if-its-ok-to-sell-your-car.shtml, accessed September 7, 2015.

42. See "What Is Free Software?" Free Software Foundation, Inc., last modified September 1, 2015, http://www.gnu.org/philosophy/free-sw.en.html, accessed September 7, 2015.

43. "Oops Landing Page," Keurig Green Mountain, Inc., http://www.keurig.com/content/oops, accessed September 7, 2015.

44. Brian Barrett, "Keurig's My K-Cup Retreat Shows We Can Beat DRM," *Wired*, May 8, 2015, http://www.wired.com/2015/05/keurig-k-cup-drm/, accessed September 7, 2015.

45. Ted Cooper, "Bad News for Keurig Green Mountain Investors: TreeHouse Foods Says Keurig 2.0 Technology Can Be Cracked," *Motley Fool*, June 23, 2014, http://www.fool.com/investing/general/2014/06/23/bad-news-for-keurig-green-mountain-investors-treeh.aspx, accessed September 7, 2015.

46. Karl Bode, "Keurig's Controversial Java 'DRM' Defeated by a Single Piece of Scotch Tape," *Techdirt*, December 11, 2014, https://www.techdirt.com/articles/20141210/07133329378/kuerigs-controversial-java-drm-defeated-single-piece-scotch-tape.shtml, accessed September 7, 2015.

47. Heather Long, "Keurig Green Mountain Gets Roasted. Stock Drops 10%," *CNN Money*, May 7, 2015, http://money.cnn.com/2015/05/06/investing/keurig-green-mountain-earnings-stock-fall/index.html, accessed September 7, 2015.

48. Samuel Gibbs, "Hackers Can Hijack Wi-Fi Hello Barbie to Spy on Your Children," *The Guardian*, November 26, 2015, http://www.theguardian.com/technology/2015/nov/26/hackers-can-hijack-wi-fi-hello-barbie-to-spy-on-your-children, accessed November 27, 2015.

49. Anjie Zheng, "VTech Has Yet to Put a Price on Hack, Chairman Says," *Wall Street Journal*, December 8, 2015, http://www.wsj.com/articles/vtech-has-yet-to-put-a-price-on-hack-chairman-says-1449556689, accessed March 13, 2016.

50. Parker Higgins, "Big Brother Is Listening: Users Need the Ability to Teach Smart TV's New Lessons," *Deeplinks* (blog), Electronic Frontier Foundation, February

11, 2015, https://www.eff.org/deeplinks/2015/02/big-brother-listening-users-need-ability-teach-smart-tvs-new-lessons, accessed September 7, 2015.

51. Jason Cipriani, "Vizio Reveals How It Secretly Tracks What You're Watching in IPO Plan," *Fortune*, July 26, 2015, http://fortune.com/2015/07/26/vizio-ipo/, accessed November 27, 2015.

52. Josh Lowensohn, "Former Apple Engineers Have Built a $1,495 Oven That Can Identify Your Food," *Verge*, June 9, 2015, http://www.theverge.com/2015/6/9/8751947/june-oven-identify-your-food, accessed September 7, 2015.

53. See Paolo Riva, Simona Sacchi, and Marco Brambilla, "Humanizing Machines: Anthropomorphization of Slot Machines Increases Gambling," *Journal of Experimental Psychology: Applied* (August 2015), doi: 10.1037/xap0000057.

54. Andrew Gregory and Graham Morrison, "Karen Sandler: Full Interview," *TuxRadar*, September 26, 2013, http://www.tuxradar.com/content/karen-sandler-full-interview, accessed September 7, 2015.

55. Emily Singer, "Getting Health Data from Inside Your Body," *MIT Technology Review*, November 22, 2011, http://www.technologyreview.com/news/426171/getting-health-data-from-inside-your-body/, accessed September 7, 2015.

56. Long Comment Regarding a Proposed Exemption under 17 U.S.C. 1201, Advanced Medical Technology Association Comments Regarding Proposed Class 27: Software—Networked Medical Devices at 2, No. 2014–07 (U.S. Copyright Office, March 27, 2015), http://copyright.gov/1201/2015/comments-032715/class%2027/AdvaMed_Class27_1201_2014.pdf, accessed September 7, 2015.

57. Biz Carson, "They Hacked Her Pancreas and Found Love along the Way," *Business Insider*, August 27, 2015, http://www.businessinsider.com/hacked-raspberry-pi-artificial-pancreas-2015-8, accessed September 7, 2015.

58. Michael Corkery and Jessica Silver-Greenberg, "Miss a Payment? Good Luck Moving That Car," *New York Times*, September 24, 2015, http://dealbook.nytimes.com/2014/09/24/miss-a-payment-good-luck-moving-that-car/, accessed September 7, 2015.

59. "Terms of Sale," Fitbit, Inc., last modified December 30, 2014, http://www.fitbit.com/legal/terms-of-sale, accessed September 7, 2015.

60. "Fitbit Privacy Policy," Fitbit, Inc., last modified December 9, 2014, http://www.fitbit.com/privacy, accessed September 7, 2015.

61. Kate Crawford, "When Fitbit Is the Expert Witness," *Atlantic*, November 19, 2014, http://www.theatlantic.com/technology/archive/2014/11/when-fitbit-is-the-expert-witness/382936/, accessed September 7, 2015.

62. "Technology Breakthrough: A Prosthetic Device That Connects to Patients and Health Care Providers via Mobile Communications Technology," *Business Wire*, September 5, 2012, http://www.businesswire.com/news/home/20120905006102/en/Technology-Breakthrough-Prosthetic-Device-Connects-Patients-Health, accessed September 7, 2015.

63. Dan Goodin, "Vast Array of Medical Devices Vulnerable to Serious Hacks, Feds Warn," *Ars Technica* (blog), June 13, 2013, http://arstechnica.com/security/2013/06/vast-array-of-medical-devices-vulnerable-to-serious-hacks-feds-warn/, accessed September 7, 2015; Jordan Robertson, "McAfee Hacker Says Medtronic Insulin Pumps Vulnerable to Attack," *Bloomberg*, February 29, 2012, http://www.bloomberg.com/news/articles/2012-02-29/mcafee-hacker-says-medtronic-insulin-pumps-vulnerable-to-attack, accessed September 7, 2015; Barnaby J. Feder, "A Heart Device Is Found Vulnerable to Hacker Attacks," *New York Times*, March 12, 2008, http://www.nytimes.com/2008/03/12/business/12heart-web.html, accessed September 7, 2015; William H. Maisel, "Semper Fidelis—Consumer Protection for Patients with Implanted Medical Devices," *New England Journal of Medicine* 358 (2008): 985–987, doi: 10.1056/NEJMp0800495.

64. Dan Goodin, "Dick Cheney Altered Implanted Heart Device to Prevent Terrorist Hack Attacks," *Ars Technica* (blog), October 19, 2013, http://arstechnica.com/security/2013/10/dick-cheney-altered-implanted-heart-device-to-prevent-terrorist-hack-attacks/, accessed September 7, 2015.

65. Joe Uchill, "What New DMCA Rules Mean for Medical Device Research," *Christian Science Monitor*, October 30, 2015, http://www.csmonitor.com/World/Passcode/2015/1030/What-new-DMCA-rules-mean-for-medical-device-research, accessed November 21, 2015.

9 Patents and the Ordinary Pursuits of Life

1. Lexmark Int'l Inc. v Static Control Components, Inc., 387 F.3d 522 (6th Cir. 2005).

2. Bloomer v. McQuewan, 55 U.S. (14 How.) 539, 549 (1852). The Court also articulated a specific economic rationale for such a limitation, one that tracks identically with the rationale for copyright exhaustion, that patentees "are entitled to but one royalty for a patented machine."

3. Ibid. at 549–550.

4. Adams v. Burke, 84 U.S. (17 Wall.) 453, 455 (1873).

5. Ibid.

6. 157 U.S. 659, 666–667 (1875).

7. Jill Jonnes, *Empires of Light: Edison, Tesla, Westinghouse, and the Race to Electrify the World* (New York: Random House, 2003), 176–190.

8. Ibid.

9. Ibid.

10. Ibid., 190.

11. Motion Picture Patents Co. v. Universal Film Mfg. Co., 243 U.S. 502, 511 (1917).

12. See Zechariah Chafee Jr., "The Music Goes Round and Round: Equitable Servitudes and Chattels," *Harvard Law Review* 69 (1956): 1250–1264, 1261; Thomas W. Merrill and Henry E. Smith, "Optimal Standardization in the Law of Property: The Numerus Clausus Principle," *Yale Law Journal* 110 (2000): 1–70, 26–28.

13. Motion Picture Patents, 243 U.S. at 519.

14. Kirtsaeng v. John Wiley & Sons, Inc., 133 S. Ct. 1351, 1363 (2013).

15. Ibid.

16. 976 F.2d 700 (Fed. Cir. 1992).

17. 304 U.S. 175 (1938).

18. Ibid. at 180.

19. 976 F.2d at 702.

20. Quanta Computer, Inc. v. LG Elecs., Inc., 553 U.S. 617, 621 (2008).

21. Ibid. at 630.

22. Ibid. at 637 n.7.

23. Mallinckrodt v. Medipart, 976 F.2d at 708 (citation omitted).

24. 264 F. 3d 1094 (Fed. Cir. 2001).

25. 133 U.S. 697 (1890).

26. Ibid. at 703.

27. See John A. Rothchild, "Exhausting Extraterritoriality," *Santa Clara Law Review* 51 (2011): 1199–1201, 1205–1206.

28. *Kirtsaeng*, 133 S. Ct. at 1371.

29. Frank R. Lichtenberg, "Pharmaceutical Price Discrimination and Social Welfare," *Capitalism and Society* 5, no. 1 (2010): 1–29, at 4.

30. See Sarah R. Wasserman Rajec, "Free Trade in Patented Goods: International Exhaustion for Patents," *Berkeley Technology Law Journal* 29 (2014): 317–376, at

361–367; Ariel Katz, "The First Sale Doctrine and the Economics of Post-Sale Restraints," *Brigham Young University Law Review* (2014): 55–142, at 80–81.

31. See Katz, "The First Sale Doctrine," 81 (citing F. M. Scherer and Jayashree Watal, "Post-TRIPS Options for Access to Patented Medicines in Developing Nations," *Journal of International Economic Law* 5 [2002]: 913).

32. See http://doggett.house.gov/images/DoggettLetter1.11.pdf, accessed April 10, 2016.

33. Liyan Chen, "Best of the Biggest: How Profitable Are the World's Largest Companies," *Forbes*, May 13, 2014, http://www.forbes.com/sites/liyanchen/2014/05/13/best-of-the-biggest-how-profitable-are-the-worlds-largest-companies/, accessed November 20, 2015.

34. Lexmark Int'l, Inc. v. Impression Products, Inc., No. 2014-1617, 2016 WL 559042 (Fed. Cir. Feb. 12, 2016).

10 Ownership's Uncertain Future

1. Richard Fry, *Young Adults After the Recession: Fewer Homes, Fewer Cars, Less Debt* (Washington, DC: Pew Research Center, 2013), 2, http://www.pewsocialtrends.org/files/2013/02/Financial_Milestones_of_Young_Adults_FINAL_2-19.pdf, accessed September 4, 2015: "The share of younger households owning their primary residence fell sharply from 40 percent in 2007 to 34 percent in 2011 ... In 2007, 73 percent of households headed by an adult younger than 25 owned or leased at least one vehicle. By 2011, 66 percent of these young households had a vehicle."

2. Tim Logan, Emily Alpert Reyes, and Ben Poston, "Airbnb and Other Short-Term Rentals Worsen Housing Shortage, Critics Say," *Los Angeles Times*, March 11, 2015, http://www.latimes.com/business/realestate/la-fi-airbnb-housing-market-20150311-story.html, accessed September 4, 2015; Laura Kusisto, "Airbnb Pushes Up Apartment Rents Slightly, Study Says," *Wall Street Journal*, March 30, 2015, http://blogs.wsj.com/developments/2015/03/30/airbnb-pushes-up-apartment-rents-slightly-study-says/, accessed September 4, 2015.

3. Rachel Monroe, "More Guests, Empty Houses," *Slate*, February 13, 2014, http://www.slate.com/articles/business/moneybox/2014/02/airbnb_gentrification_how_the_sharing_economy_drives_up_housing_prices.html, accessed September 4, 2015.

4. Stephen Gandel, "Uber-nomics: Here's What It Would Cost Uber to Pay Its Drivers as Employees," *Fortune*, September 17, 2015, http://fortune.com/2015/09/17/ubernomics/, accessed November 20, 2015.

5. Alison Griswold, "Uber Surged Prices during the Sydney Hostage Crisis. It Needs to Do Better," *Moneybox* (blog), *Slate*, December 15, 2014, http://www.slate.com/

blogs/moneybox/2014/12/15/uber_sydney_hostage_crisis_it_s_time_for_uber_to_re_evaluate_how_it_prices.html, accessed September 4, 2015.

6. Dan Kedmey, "This Is How Uber's 'Surge Pricing' Works," *Time*, December 15, 2014, http://time.com/3633469/uber-surge-pricing/, accessed September 4, 2015.

7. "Random House Launches New Digital-Only Imprints," *Publisher's Weekly*, November 29, 2012, http://www.publishersweekly.com/pw/by-topic/digital/content-and-e-books/article/54930-random-house-launches-new-digital-only-imprints.html, accessed September 4, 2015.

8. Vish Khanna, "Why Aren't There More Vinyl Pressing Plants?" *The Pitch* (blog), *Pitchfork*, June 9, 2014, http://pitchfork.com/thepitch/363-vinyl-shortage/, accessed September 4, 2015.

9. Ted Sarandos, "Why You'll See Some High Profile Movies Leave Netflix US Next Month," *Netflix US & Canada Blog*, August 30, 2015, https://web.archive.org/web/20151126144555/http://blog.netflix.com/2015/08/why-youll-see-some-high-profile-movies.html (site discontinued), accessed November 20, 2015.

10. If this seems farfetched, consider the statement of Ford's VP of marketing and sales: "We know everyone who breaks the law, we know when you're doing it. We have GPS in your car, so we know what you're doing." Jim Edwards, "Ford Exec Retracts Statements about Tracking Drivers with the GPS in Their Cars," *Business Insider*, January 9, 2014, http://www.businessinsider.com/ford-jim-farley-retracts-statements-tracking-drivers-gps-2014-1, accessed September 4, 2015.

11. See Jessica Litman, "Revising Copyright Law for the Information Age," *Oregon Law Review* 75 (1996): 19–48 (describing a legislative process dominated by industry insiders).

12. Maria A. Pallente, "The Next Great Copyright Act" (lecture, Columbia University, New York, March 4, 2013), https://web.law.columbia.edu/sites/default/files/microsites/kernochan/files/Pallante-The%20Next%20Great%20Copyright%20Act.pdf, accessed September 4, 2015.

13. Lesley Fair, "Full Disclosure," *FTC Business Blog*, September 23, 2014, https://www.ftc.gov/news-events/blogs/business-blog/2014/09/full-disclosure, accessed September 4, 2015.

14. Ibid.; Nancy S. Kim, *Wrap Contracts* (New York: Oxford University Press, 2013), 89–90.

15. Council Directive 93/13 1993 O.J. (l. 95); Jane K. Winn and Mark Webber, "The Impact of EU Unfair Contract Terms Law on US Business-to-Consumer Internet Merchants," *Business Lawyer* 62 (November 2006): 209–228.

16. Rice University, "Removal of Restrictions Can Decrease Music Piracy, Study Suggest," *ScienceDaily*, October 10, 2011, http://www.sciencedaily.com/releases/2011/10/111007113944.htm, accessed September 4, 2015.

17. One practical problem is that the United States is a signatory to the WIPO Copyright Treaty. That treaty requires parties to "provide adequate legal protection and effective legal remedies against the circumvention of effective technological measures." WIPO Copyright Treaty art. 11, April 12, 1997, S. Treaty Doc. No. 105-17 (1997). Conceivably, the United States could limit anti-circumvention law in a manner consistent with exhaustion since the uses owners would make would be "permitted by law." Ibid.

18. Chamberlain Grp., Inc. v. Skylink Techs., Inc., 381 F.3d 1178, 1202 (Fed. Cir. 2004).

19. See Storage Tech. Corp. v. Custom Hardware Eng'g & Consulting, Inc., 421 F.3d 1307, 1319 (Fed. Cir. 2005). The Fifth Circuit rejected an "interpretation [that] would permit liability under section 1201(a) for accessing a work simply to view it or to use it within the purview of 'fair use' permitted under the Copyright Act." MGE UPS Sys. Inc. v. GE Consumer & Indus., 612 F.3d 760, 765 (5th Cir. 2005).

20. Although outside the scope of this book, in some instances breaking DRM could violate the Computer Fraud and Abuse Act. See 18 U.S.C. § 1030 (creating civil and criminal liability for "knowingly access[ing] a computer without authorization or exceeding authorized access").

21. H.R. 3048, 105th Cong. (1997).

22. Christopher Wolf, *The Digital Millennium Copyright Act: Text, History, and Caselaw* (Silver Spring, MD: Pike & Fischer, 2003), 460.

23. H.R. 862, 114th Cong. (2015).

24. "Grassley & Leahy Call for Copyright Study," Sen. Chuck Grassley's webpage, October 22, 2015, http://www.grassley.senate.gov/news/news-releases/grassley-leahy-call-copyright-study, accessed November 20, 2015.

25. See 17 U.S.C. § 107.

26. Pamela Samuelson, "Unbundling Fair Uses," *Fordham Law Review* 77 (April 2009): 2537–2621, at 2602–2615; Michael J. Madison, "A Pattern-Oriented Approach to Fair Use," *William & Mary Law Review* 45, no. 4 (2004): 1525–1687, at 1687.

27. See 17 U.S.C. §§ 109 & 117.

28. Platt & Munk Co. v. Republic Graphics, Inc., 315 F.2d 847, 854 (2d Cir. 1963) (noting that "the ultimate question embodied in the 'first sale' doctrine is 'whether or not there has been such a disposition of the article that it may fairly be said that

the patentee [or copyright proprietor] has received his reward for the use of the article'" (alteration in original) (quoting United States v. Masonite Corp., 316 U.S. 265, 278 [1942]); Parfums Givenchy, Inc. v. C & C Beauty Sales, Inc., 832 F. Supp. 1378, 1389 (C.D. Cal. 1993) ("The distribution right and the first sale doctrine rest on the principle that the copyright owner is entitled to realize no more and no less than the full value of each copy or phonorecord upon its disposition"); Burke & Van Heusen, Inc. v. Arrow Drug, Inc., 233 F. Supp. 881, 884 (E.D. Pa. 1964) ("The ultimate question under the 'first sale' doctrine is whether or not there has been such a disposition of the copyrighted article that it may fairly be said that the copyright proprietor has received his reward for its use.").

29. Christina Mulligan, "A Numerus Clausus Principle for Intellectual Property," *Tennessee Law Review* 80 (2013), 280: "Digital works have greater hurdles to preservation than analog or physical copies of works; in addition to preserving a copy and translating the language, 'digital translation' presents an additional problem"; Reese, "The First Sale Doctrine in the Era of Digital Networks," *Boston College Law Review* 44 (2003): 577–652, at 633–639 (noting the difficulties of preserving readable digital content in an environment in which languages, software, hardware, and file formats undergo rapid change).

30. 17 U.S.C. § 107.

31. 17 U.S.C. § 109(b).

32. Tiffany (NJ) Inc. v. eBay, Inc., 600 F.3d 93 (2d Cir. 2010), *aff'g* 576 F. Supp. 2d 463 (S.D.N.Y. 2008).

33. Viacom Int'l, Inc. v. YouTube, Inc., 676 F.3d 19 (2d Cir. 2012), *aff'g in part, rev'g and rem'g in part,* 718 F. Supp. 2d 514 (S.D.N.Y. 2010).

34. Case C-128/11, UsedSoft GmbH v. Oracle Int'l Corp., 2012 E.C.R. I-0000.

35. Rb. Den Haag, 3 september 2014, ECLI:NL:RBDHA:2014:10962, *IEF* 14164 (*VOB/Stichting Leenrecht e.a.*) (Neth.), http://uitspraken.rechtspraak.nl/inziendocument?id=ECLI:NL:RBDHA:2014:10962, accessed September 4, 2015.

36. Hof Amsterdam, 20 januari 2015, ECLI:NL:GHAMS:2015:66, *NUV/Tom Kabinet* (Neth.), http://uitspraken.rechtspraak.nl/inziendocument?id=ECLI:NL:GHAMS:2015:66, accessed September 4, 2015.

37. Ibid.

38. Christina Mulligan, "Killing Copyright," on file with authors.

39. *Public Hearing Filed in Response to 65 FR 63626,* November 29, 2000 (testimony of Susan Mann, National Music Publishers Association).

40. National Telecommunications & Information Administration, *Report to Congress: Study Examining 17 U.S.C. Sections 109 and 117 Pursuant to Section 104 of the Digital Millennium Copyright Act* (2001), https://www.ntia.doc.gov/report/2001/report-congress-study-examining-17-usc-sections-109-and-117-pursuant-section-104-digital, accessed September 4, 2015.

41. U.S. Copyright Office, *DMCA Section 104 Report* (2001), http://www.copyright.gov/reports/studies/dmca/sec-104-report-vol-1.pdf, accessed September 4, 2015.

42. The proposed DPP standard would allow consumers to share digital files with anyone they chose using a "give" button. But when a recipient clicked a "take" button, access to all other copies originating with the initial purchaser would be disabled. Paul Sweazey, "Introduction to Digital Personal Property," in *Consumers in the Information Society: Access, Fairness and Representation*, ed. Jeremy Malcolm (Kuala Lumpur: Consumers International, 2012), 53–71.

43. Erich Ringewald, Secondary market for digital objects, US Patent 8,364,595, filed May 5, 2009, issued January 29, 2013.

44. Eliza C. Block and Marcel van Os, Managing access to digital content items, US Patent Application 20130060616, filed June 22, 2012.

45. Jack Bertram Coronel and Joseph R. Coronel, Method, system, and device for providing a market for digital goods, US Patent 8,631,505, filed March 16, 2013, issued January 14, 2014; Brian K. Buchheit. Secondary marketplace for digital media content, US Patent 8,359,246, filed March 19, 2010, issued January 22, 2013.

46. Block and van Os, Managing access to digital content items: "A portion of the proceeds of the 'resale' may be paid to the creator of the digital content item."

47. S. 2045, 113th Cong. (2014); H.R. 4103, 113th Cong. (2014).

48. See Sam Francis Foundation v. Christies, 78 F.3d 1320 (9th Cir. 2015).

49. Guy A. Rub, "The Unconvincing Case for Resale Royalties," *Yale Law Journal Forum* 124 (2014): 1, http://www.yalelawjournal.org/forum/the-unconvincing-case-for-resale-royalties, accessed September 4, 2015.

50. For a thorough discussion of the insights bitcoin offers for property in general and digital assets in particular, see Joshua A. T. Fairfield, "Bitproperty," *Southern California Law Review* 88 (May 2015): 805–874.

51. James Grimmelmann and Arvind Narayanan, "The Blockchain Gang," *Slate*, February 16, 2016, http://www.slate.com/articles/technology/future_tense/2016/02/bitcoin_s_blockchain_technology_won_t_change_everything.html, accessed April 10, 2015.

52. Marc Andreessen, "Why Bitcoin Matters," *Dealbook* (blog), *New York Times*, January 21, 2014, http://dealbook.nytimes.com/2014/01/21/why-bitcoin-matters/, accessed September 4, 2015.

53. Although the public ledger does not include sender and recipient names, it does include account numbers. There are steps users can take to maintain privacy, but they are admittedly imperfect. But efforts are underway for new privacy-enhancing tools. See Bitcoin, "Protecting Your Privacy," https://bitcoin.org/en/protect-your-privacy, accessed September 4, 2015.

54. As with any property, the value of a digital asset can fluctuate. Dedicated Destiny players will tell you that the once-mighty Gjallarhorn was "nerfed" by game maker Bungie, rendering it far less coveted.

55. Fairfield, "Bitproperty."

56. See Arnold S. Weinrib, "Information and Property," *University of Toronto Law Journal* 38 (1988): 117–150, at 120: "It also makes plain the conclusory nature of the term 'property': it is a legal characterization, a statement that the court has chosen to assign a particular form of protection to the interest in question" (footnote omitted); see also Johnson v. M'Intosh, 21 U.S. (8 Wheat.) 543, 572 (1823) (stating that property claims "must be admitted to depend entirely on the law of the nation in which they lie"); Jeremy Bentham, *The Theory of Legislation*, ed. C. K. Ogden (London: Oxford University Press, 1950), 113: "Property and law are born together, and die together. Before laws were made there was no property; take away laws, and property ceases."

Index